Lecture Notes in Business Information Processing 546

LNBIP reports state-of-the-art results in areas related to business information systems and industrial application software development – timely, at a high level, and in both printed and electronic form.

The type of material published includes

- Proceedings (published in time for the respective event)
- Postproceedings (consisting of thoroughly revised and/or extended final papers)
- Other edited monographs (such as, for example, project reports or invited volumes)
- Tutorials (coherently integrated collections of lectures given at advanced courses, seminars, schools, etc.)
- Award-winning or exceptional theses

LNBIP is abstracted/indexed in DBLP, EI and Scopus. LNBIP volumes are also submitted for the inclusion in ISI Proceedings.

Boris Delibašić · Sandro Radovanović ·
Pavlos Delias · Shaofeng Liu ·
Sérgio Pedro Duarte
Editors

Decision Support Systems XIV

Human-Centric Group Decision, Negotiation and Decision Support Systems for Societal Transitions

11th International Conference on Decision Support System Technology, ICDSST 2025
Belgrade, Serbia, May 26–29, 2025, Proceedings

Springer

Editors
Boris Delibašić
University of Belgrade
Belgrade, Serbia

Sandro Radovanović
University of Belgrade
Belgrade, Serbia

Pavlos Delias
University of Thrace
Xanthi, Greece

Shaofeng Liu
University of Plymouth
Plymouth, Devon, UK

Sérgio Pedro Duarte
University of Porto
Porto, Portugal

ISSN 1865-1348 ISSN 1865-1356 (electronic)
Lecture Notes in Business Information Processing
ISBN 978-3-031-90862-0 ISBN 978-3-031-90863-7 (eBook)
https://doi.org/10.1007/978-3-031-90863-7

This Springer imprint is published by the registered company Springer Nature Switzerland AG
The registered company address is: Gewerbestrasse 11, 6330 Cham, Switzerland

Preface

This eleventh edition of the EWG-DSS Decision Support Systems proceedings, published in the LNBIP series, presents a selection of high-quality papers from the 11th International Conference on Decision Support System Technology (ICDSST 2025), held in May 2025 in Belgrade, Serbia. The conference was organized by the University of Belgrade, Faculty of Organizational Sciences (Serbia) in collaboration with the EURO Working Group on Decision Support Systems (EWG-DSS). The ICDSST series of conferences serves as a vital platform for the European and international DSS communities to exchange ideas, share innovative solutions, and explore emerging trends in decision-making support. As such, the conference brought together researchers and practitioners from academia and industry to discuss advancements in decision support system technologies.

This year's conference focused on "Decision Support System Technology in the AI Era", emphasizing the transformative role of artificial intelligence in decision-making frameworks in diverse domains. Rapid advances in AI have presented new challenges and opportunities in DSS development, influencing areas such as machine learning-driven analytics, real-time decision making, and explainable AI models. In recent years, AI-driven decision support has gained unprecedented relevance as organizations and policy makers strive to improve efficiency, precision, and transparency in complex decision-making processes. The proliferation of big data, deep learning, and generative AI models has enabled DSS to process vast amounts of information, derive actionable insights, and adapt to dynamic environments in previously unimaginable ways. However, these advancements also bring challenges related to interpretability, accountability, and ethical considerations, necessitating ongoing research and dialogue. This makes "Decision Support System Technology in the AI Era" a highly interesting topic for discussion, as it directly impacts industries, governments, and society at large. This volume provides valuable insights into the latest advancements in decision support systems, mainly highlighting developments in real-world applications. We hope that the findings and discussions presented in these proceedings will inspire further research and collaboration within the DSS community, fostering the development of next-generation decision support technologies.

A total of 39 submissions were received for ICDSST 2025, of which 10 full papers were carefully selected through a double-blind peer review process. Each paper had two reviewers and a meta-reviewer who controlled the process and suggested a final decision. In total, ICDSST 2025 had 44 reviewers, mostly from Europe, but also from North and South America and Africa. These papers have been organized into three thematic sections: Decision Support Systems, Artificial Intelligence and Machine Learning, and Decision Support System Challenges. The selected contributions represent a broad spectrum of innovative research, including AI-powered decision support models, novel methodologies to enhance decision-making processes, and discussions on the ethical and practical implications of AI integration in DSS.

The Decision Support Systems section explores various decision-making frameworks and methodologies across industries. The selected articles examine diverse applications, including a technology organization-environment decision framework that investigates the adoption of emerging technology on farms in China, the role of peer learning cycles in supporting decision making for food-sharing organizations, decision support mechanisms within Brazil's Plan for Control of Waste and Contaminants in the dairy industry, and the application of blockchain technology in maritime logistics, particularly to address the tactical berth allocation problem.

The Artificial Intelligence and Machine Learning section highlights how AI technologies enhance decision support capabilities. The research in this section includes an analysis of interface preferences in Decision Support Systems using eye tracking, a study on the role of AI and generative AI in e-learning processes and outcomes from a system-wide perspective, an exploration of how AI contributes to improved decision making in small and medium enterprises, and a sentiment classification study focused on product influencers.

Lastly, the section on Decision Support System Challenges addresses emerging difficulties and complexities in decision-making environments. The included studies investigate a system dynamics model designed to raise awareness of green transition challenges in higher education institutions, as well as a framework for proactive debt management in business process management, shifting the focus from technical debt to broader business process debt considerations.

We express our gratitude to all the authors, reviewers, and members of the organizing committee who contributed to the success of ICDSST 2025. Their efforts ensured the high quality of this publication and of the conference as a whole. Special thanks go to our keynote speakers for sharing their expertise and vision on the future of decision support systems in the AI era. We also appreciate the support of our institutional and industry partners who helped make this event possible.

We hope that you find this volume both insightful and inspiring.

March 2025

Boris Delibašić
Sandro Radovanović
Pavlos Delias
Shaofeng Liu
Sérgio Pedro Duarte

Organization

Honorary Chairs

Fatima Dargam	SimTech-Reach Consulting, Austria
Pascale Zarate	Université Toulouse Capitole-IRIT, France

Conference Chairs

Boris Delibašić	University of Belgrade, Serbia
Sandro Radovanović	University of Belgrade, Serbia

Program Committee Chairs

Ana Paula Cabral Seixas Costa	Federal University of Pernambuco, Brazil
Pavlos Delias	Kavala Institute of Technology, Greece
Boris Delibašić	University of Belgrade, Serbia
Sérgio Pedro Duarte	University of Porto, Portugal
María Teresa Escobar	University of Zaragoza, Spain
Shaofeng Liu	University of Plymouth, UK
Jason Papathanasiou	University of Macedonia, Greece
Sandro Radovanović	University of Belgrade, Serbia

Organizing Committee

Miloš Jovanović	University of Belgrade, Serbia
Andrija Petrović	University of Belgrade, Serbia
Milan Vukićević	University of Belgrade, Serbia

Program Committee

Adiel Teixeira de Almeida	Federal University of Pernambuco, Brazil
Md Asaduzzaman	Staffordshire University, UK
Dragana Bečejski-Vujaklija	Serbian Society for Informatics, Serbia
Marko Bohanec	Jožef Stefan Institute, Slovenia

Ana Paula Cabral Seixas Costa	Federal University of Pernambuco, Brazil
Boris Delibašić	University of Belgrade, Serbia
Pavlos Delias	Kavala Institute of Technology, Greece
Panagiota Digkoglou	University of Macedonia, Greece
Sérgio Pedro Duarte	University of Porto, Portugal
María Teresa Escobar	University of Zaragoza, Spain
Sean Eom	Southeast Missouri State University, USA
Fátima Dargam	SimTech-Reach Consulting, Austria
Miloš Jovanović	University of Belgrade, Serbia
Daouda Kamissoko	IMT Mines Albi, France
Kathrin Kirchner	Technical University of Denmark, Denmark
Marc Kilgour	Wilfrid Laurier University, Canada
Antonio Lobo	Universidade do Porto, Porto
João Lourenço	Universidade de Lisboa, Portugal
Isabelle Linden	University of Namur, Belgium
Shaofeng Liu	University of Plymouth, UK
Francisco Marmolejo-Cossio	Boston College, USA
José Maria Moreno-Jiménez	University of Zaragoza, Spain
Ben C. K. Ngan	Worcester Polytechnic Institute, USA
Festus Oderanti	Liberty University, USA
Femi Olan	University of Essex, UK
Jason Papathanasiou	University of Macedonia, Greece
Andrija Petrović	University of Belgrade, Serbia
Sandro Radovanović	University of Belgrade, Serbia
Rita Ribeiro	UNINOVA – CA3, Portugal
Alexander Smirnov	Russian Academy of Sciences, Russia
Karim Soliman	Arab Academy for Science, Technology, and Maritime Transport, Egypt
Jorge Freire de Sousa	University of Porto, Portugal
Vlasta Sikimić	Eindhoven University of Technology, Netherlands
Alexis Tsoukias	Université Paris Dauphine, France
George Tsaples	University of Macedonia, Greece
Fernando Tricas Garcia	University of Zaragoza, Spain
Alberto Turón	University of Zaragoza, Spain
Konstantinos Vergidis	University of Macedonia, Greece
Rudolf Vetschera	University of Vienna, Austria
Milan Vukićević	University of Belgrade, Serbia
Andy Wong	University of Strathclyde, UK
Uchitha Jayawickrama	Loughborough University, UK
Pascale Zaraté	Université Toulouse Capitole-IRIT, France
Guoqing Zhao	Swansea University, UK

Poster Session Committee Members

Naveen Achyuta	Amazon, USA
Manan Buddhadev	IEEE, USA
Aditya Gupta	Amazon, USA
Balaji Ingole	Gainwell Technologies LLC, USA
Mohit Menghnani	Roblox, USA
Sandeep Ravindra Tengali	Atlassian, USA
Jayant Tyagi	Salesforce, USA

Sponsors

Working Group on Decision Support Systems (https://ewgdss.wordpress.com)
Association of European Operational Research Societies (www.euro-online.org)

Institutional Sponsors

Faculty of Organizational Sciences
University of Belgrade, Serbia
(http://www.fon.bg.ac.rs/eng/)
The Ministry of Science, Technological Development and Innovation of the Republic of Serbia
(https://nitra.gov.rs/en)
Graduate School of Management, Faculty of Business
University of Plymouth, UK
(http://www.plymouth.ac.uk/)
School of Management Science and Accounting
Democritus University of Thrace, Greece
(https://www.duth.gr)
Federal University of Pernambuco, Brazil
(https://www.ufpe.br/inicio)
University of Macedonia, Department of Business Administration
Thessaloniki, Greece
(http://www.uom.gr/index.php?newlang=eng)

Contents

Decision Support System Challenges

Decision Support Systems

A Technology-Organization-Environment Decision Framework for Understanding Emerging Technology Adoption on Farms: An Empirical Investigation from China

Guoqing Zhao[1(✉)], Xiaoning Chen[2], Jiandong Hu[3], and Agung Sutrisno[4]

[1] School of Management, Swansea University, Swansea, UK
guoqing.zhao@swansea.ac.uk

[2] Plymouth Business School, University of Plymouth, Plymouth, UK
xiaoning.chen@plymouth.ac.uk

[3] College of Mechanical and Electronical Engineering, Henan Agricultural University, Zhengzhou, China
jdhu@henau.edu.cn

[4] Department of Mechanical Engineering, Sam Ratulangi University, Manado, Indonesia
agungsutrisno@unsrat.ac.id

Abstract. Emerging technologies (ETs) are being applied on farms to promote efficiency, productivity and automation. However, their low maturity, small-scale application and high investment costs impede further implementation in more farming areas. The total factor productivity (TFP) associated with China's agricultural mechanization rate has grown exponentially, reaching 71.25% in 2023, largely due to applications of ETs on farms. Thus, grounded in a technology-organization-environment (TOE) framework, we develop an inductive, qualitative approach to explore factors affecting Chinese farms' ET adoption. We conducted in-depth interviews with knowledgeable farmers who have applied ETs, followed by thematic analysis to identify factors. Our findings make novel theoretical and managerial contributions. First, we identify 14 factors facilitating adoption of ETs on farms, most of which have seldom been mentioned by previous scholars, such as the hierarchical cultural environment, wide deployment of digital platforms to disseminate information, increasing farm size and development of high-quality farmland. Second, the technological, organizational and environmental dimensions collectively determine successful adoption of ETs. The environmental dimension is responsible for providing resources (e.g., technology, policy and finance) to develop ETs, disseminating information and fostering competition between farms. The organizational dimension prepares farms to apply ETs, and the technological dimension is responsible for developing reliable and easy-to-use technologies. Based on the TOE decision framework, our findings have managerial implications for farmers, suggesting that increasing farm size (e.g., more than 200 acres of farmland) may facilitate ET deployment.

Keywords: Emerging technologies · farms · China's agri-food industry · technology-organization-environment framework

B. Delibašić et al. (Eds.): ICDSST 2025, LNBIP 546, pp. 3–14, 2025.
https://doi.org/10.1007/978-3-031-90863-7_1

1 Introduction

The agricultural industry plays crucial roles in feeding the world's growing population, contributing to regional and national economic growth, shaping social structures and preserving the cultural practices, traditions and heritage of many communities. Its prosperity is widely considered critical to tackling the sustainable development goals of zero poverty, zero hunger and climate action [1]. However, the industry faces various challenges to achieving sustainable development. For example, climate change-triggered extreme weather conditions (e.g., droughts, floods and record-breaking heatwaves) have huge impacts on food production, and significantly limiting the availability and accessibility of resources. Rapid urbanization over the last six decades has led to increasing numbers of people living in cities and metropolitan areas, as well as a significant reduction in agricultural land. Currently, only 12% of the global land surface is used for crop production, and continued decline is expected owing to rapid urban expansion [2]. Extensive loss of biological diversity is making food supplies more vulnerable to pests and diseases, resulting in more people supplementing their diets with industrially processed foods. Other challenges frequently mentioned by scholars are population growth, soil erosion, intensive agriculture, low investment in agriculture and rising demand for higher-quality food [3, 4]. As a result of expected growth in population and income levels, the environmental effects of the agricultural sector may increase by 50%–90% without technological innovations and dedicated mitigation measures, thus reaching planetary boundaries by 2050 [5].

One feasible way to tackle these agricultural challenges is to apply emerging technologies (ETs), referring to new and rapidly developing technologies with the potential to significantly change various industries [5]. For example, solutions integrating the internet-of-things (IoT), blockchain and satellite imagery enable data sharing between different devices, and thus contribute to enhanced transparency and improved resource management [7]. Artificial intelligence (AI), computer vision and deep learning-enabled robots can be combined to monitor soil health and precisely track and predict the weather. Big data analytics (BDA) are effectively used to improve farming operations, for example with enhanced agricultural skills and weather forecasts, by analyzing massive pools of agricultural data [8]. ETs promise various benefits enabling farmers to build highly efficient, automated farm, but their adoption rate remains relatively low and fragmented. This is because adopting ETs on farms requires various enabling conditions to ensure successful implementation, integration and utilization. For example, [9] state that intentions to use and facilitating conditions may influence adoption of ETs on farms. More specifically, they highlight effort expectancy, performance expectancy and social influence as critical factors. [10]'s review construes that factors affect ET adoption on small-scale farms can be categorized into economic, political, social, and governance and institutional. Existing studies have extensively explored and discussed factors affecting ET adoption in the agricultural sectors of various countries (e.g., South Korea and South Africa), but deeper understanding is impeded by the fragmented, unexplanatory and atheoretical nature of much extant literature. Recent reviews of the adoption of emerging farming technologies also highlight a dearth of empirical studies focusing specifically on farm-level ET adoption processes [11, 12].

This study fills this gap by exploring factors affecting ET adoption on Chinese farms. We formulated two research questions to gain an in-depth understanding of the phenomenon. RQ1: What factors facilitate adoption of ETs on farms? RQ2: How are these factors connected to achieve synergies? To answer the first question, we collected data through semi-structured interviews with Chinese farmers, followed by thematic analysis to identify factors affecting ET adoption of farms. To answer the second question, grounded in a technology-organization-environment (TOE) framework, we thoroughly examined the factors and explored how they can be connected to achieve synergies.

In the remainder of this paper, in Sect. 2 we review existing literature on factors affecting ET adoption on farms and identify research gaps. In Sect. 3 we introduce and justify our adoption of the TOE framework, semi-structured interviews and thematic analysis. In Sect. 4 we present and discuss the identified factors, and explain our unique theoretical and managerial contributions. Finally, in Sect. 5 we draw some conclusions.

2 Literature Review

In this section, we review the factors, conditions, decision frameworks, technologies and theories applied in relation to adoption of ETs on farms, leading to identification of research gaps.

2.1 ETs: Definitions and Applications on Farms

Extensive scholarly discussions have provided rich insights into ETs. For example, [19] define ETs as technologies that have high potential but have not demonstrated their value or settled down into any kind of consensus. [20]'s definition is similar, proposing that ETs are high potential technologies that may gain social relevance in the next 10 to 15 years. According to [21], ETs are characterized by novelty and growth. These definitions are too vague and lacking consensus for use in this study. Instead, we use [22]'s definition, which highlights five characteristics of ETs: radical novelty, relatively fast growth, coherence, prominent impact, and uncertainty and ambiguity.

We began by extensively searching relevant literature and consulting experts. Farmers adopt ETs to support two areas of development: food and crop production, and management activities [23]. For example, [24]'s review of agriculture 4.0 development reveals that ETs are deployed to support various stages of agricultural production. In the pre-field stage, AI and next-generation gene-editing technology can be used to support genetics development, while digital twins, 3D food printing and sensing technologies are widely used to develop new seeds. In the in-field stage, various ETs are used to support planting and harvesting, including cloud computing, mobile and autonomous robots, intelligent greenhouses and electrical agricultural machinery. Finally, in the post-field stage, ETs such as BDA, AI, cybersecurity, intelligent software algorithms and machine learning are being developed to support distribution, processing and consumption processes. [25] propose that IoT, BDA, AI, blockchain technology, robots, augmented reality, cloud technology and virtual reality can be used on farms to measure soil nutrition and monitor crop health. Scholarly consensus is lacking on ETs used on farms because agriculture is a broad subject and ET adoption must take account of specific conditions such as crop and soil types.

2.2 Adoption of ETs on Farms: Enabling Factors and Conditions

Scholars have conducted extensive analysis of factors affecting adoption of ETs on farms. For example, [13] identify four categories of factors governing precision agriculture adoption among small-scale farmers: economic, technological, social and environmental. Amongst these, social dynamic are the most critical factor, shaped by awareness levels, knowledge dissemination pathways and entrenched cultural norms. [14] investigate critical factors affecting digital agri-tech adoption on Australian farms. They indicate that technical, discursive and social factors are most influential in digital agri-tech adoption. [15]'s review shows that 20 factors may influence adoption of digital agricultural technologies, such as facilitating conditions (ease of use), perceived usefulness, technological infrastructure, and economical and financial conditions. [16] conduct several case studies based on data collected from Italian field crop farms. Their results indicate that policymakers should pay particular attention to three enablers: farmers' digital skills, data management practices and interoperability of digital solutions. [17] explore drivers of digitalization in rural areas based on experts' views, showing that practical demands, cultural tendencies, service quality, market demands, business needs, impact reduction and control and regulatory restrictions are the main drivers. Finally, [18] confirm that subjective norms, perceived behaviour control and farmers' perceptions of usefulness impact favourably on the uptake of digital extension services.

2.3 Empirical Investigations, Theories Adopted and Countries of Focus

Many empirical studies have investigated adoption of ETs in the agricultural sector. For example, [26] explore drivers of and challenges to applying ETs in the Italian food processing industry through a case study. Driving forces include digitalization and vertical integration, robotization and interconnection of machines, and horizontal integration. [27] also employ a case study approach and collect data from an agri-food company in the sub-Saharan area. They identify 12 barriers impeding adoption of ETs, including insufficient data, lack of government incentives and unavailability of young people. [28] untangle relationships between different enablers to achieve zero hunger through IoT and blockchain technology to develop a decision support framework. Enablers such as food safety, logistics efficiency and waste reduction are included in their analysis. These studies enhance scholars' and practitioners' knowledge of ET adoption in the agri-food industry from different perspectives, but their atheoretical approach impedes further understanding. Complementing previous atheoretical empirical research, [29] employ a multiple case study grounded in middle-range theory to investigate drivers of ET use to achieve AFSC sustainability. They indicate that successful adoption of I4.0 technologies depends on social, economic and environmental forces. Notable drivers include improving working conditions, reducing carbon emissions and reducing human exposure to pesticides. Through the theoretical lens of the TOE framework and drawing on force field theory, [30] use data collected from agri-food industry practitioners to examine 22 barriers to blockchain adoption to achieve sustainable supply chains, including lack of government policies, lack of knowledge and expertise, and technological immaturity.

Table 1 summarizes other empirical studies focusing on ET adoption by the agri-food industry.

Table 1. Other empirical studies focusing on ET adoption

Source	Theory adopted and country	Topic explored
[31]	Dynamic capabilities, Italy	Blockchain adoption enablers in wineries
[32]	TOE framework and UTAUT model, Brazil	Factors affecting IoT adoption on vegetable farms
[33]	No theory adopted, India	3D printing adoption as enabler of food processing
[34]	UTAUT model, Australia	IoT adoption as enabler in the food and beverage industry
[35]	No theory adopted, India	Blockchain and IoT adoption barriers in AFSC

2.4 Research Gaps

Based on our review, we identify three research gaps. First, theories are seldom used to explore enablers of or barriers to adoption of ETs by the agri-food industry. This may be because their adoption requires contributions from individuals, organizations, governments, and even different aspects of society at large. As a result, widely used theories such as dynamic capabilities and unified theory of acceptance and use of technology (UTAUT) models are insufficient to capture all factors. Our study fills this gap by adopting a TOE framework to explain various enablers influencing adoption and use of ETs. We focus on enablers because their identification will help practitioners to overcome barriers and accelerate technology adoption.

Second, research has been conducted in various countries (e.g., Australia, Brazil, India, Italy, and South Korea) to identify and analyze of enablers or barriers to adoption of ETs, but China has received relatively little attention. China's agricultural production has increased significantly in the last decade with the help of agricultural machineries and ETs, so what makes its agricultural production successful? Our exploration of enablers of ET adoption on Chinese farms promises rich insights and potentially useful managerial contributions.

Finally, scholarly attention has focused mainly on the agri-food processing industry to explore the ET adoption process, whereas agricultural production remains relatively neglected. For example, [33] collected data from Indian food processing companies, and [34] explored IoT adoption enablers based on data collected from organizations in the food and beverage industry, rather than focusing purely on farms. To fill this gap, we conducted in-depth interviews with experienced practitioners from three Chinese farms.

3 Research Methodology

In this section, we explain our data collection and analysis methods to justify our sound research methodology.

3.1 TOE Framework

The TOE framework is a theoretical framework widely used to explain how adoption and use of new technologies are influenced by various factors, including characteristics of the technology itself, the organizational context in which it is used, and the external environment in which the organization operates [50]. We adopted the TOE framework for this study for several reasons. First, it is a well-developed theory and has been widely used to explore various factors affecting different types of technology adoption. For example, [51] employ a TOE framework to explore factors affecting adoption of smart technologies in Korean farms. [52] investigate the key determinants of adopting AI in agriculture by using a TOE framework. Its wide applicability gave us some confidence in applying it in this study. Second, the framework is clear and flexible and can be adapted to different qualitative and quantitative approaches.

3.2 Semi-structured Interviews and Empirical Data Collection

Semi-structured interviews are a widely used qualitative data collection method in which a pre-determined set of open-ended research questions is used to explore a particular theme [36]. They are adopted for several reasons. First, they have been used successfully in previous studies to collect qualitative data from agri-food industry practitioners. For example, [37] use semi-structured interviews to collect data from AFSC practitioners to investigate factors affecting cross-boundary knowledge mobilization of AFSCs. Second, semi-structured interviews provide sufficient opportunities for participants to discuss a potential topic of interest, allowing previously unknown information to emerge. They have advantages over questionnaires because people tend to take less time to complete questionnaires, which may result in low-quality information [38]. Third, agri-food industry practitioners prefer to be interviewed rather than filling in questionnaires, as evidenced by [39].

We developed an interview guide by conducting a roundtable discussion with two professors in business and management and evaluating the results with three Chinese farmers. The evaluators' comments were minor and related mainly to grammatical issues. We then established two criteria for recruiting participants with appropriate professional knowledge and working experience [40]: (1) the selected farmers should be directly involved in using ETs; and (2) they must have at least 10 years' working experience in the farming field, to ensure that they had high levels of skills, experience or expertise. Accordingly, we selected 10 Chinese farmers for interview. Each interview was recorded with permission and lasted between 45 and 60 min, and interviewees were encouraged to express their ideas freely about the topic being discussed.

3.3 Thematic Analysis and Processes

Thematic analysis is an easily applied, widely adopted and foundational method of qualitative data analysis [41]. We employed it in this study for two reasons. First, it is a highly flexible approach that can be modified to satisfy the needs of particular studies and can be used to analyze large qualitative datasets and produce rich and detailed insights. Second, the results of thematic analysis are easily understood by less educated practitioners, making it appropriate for our discussion of impacts among farmers.

We followed the four-step process proposed by [42] to implement thematic analysis. First, we transcribed and immersively read the transcripts several times to familiarize ourselves with the data. Second, we coded the data by highlighting sentences and paragraphs relevant to the topic. NVivo 13 was used to support the open coding process. In the next step of searching for and confirming themes, we categorize codes with similar meanings, checked previous literature focusing on ET adoption on farms, synthesized themes to represent these codes, and aggregated the themes into different dimensions. Finally, we used [43]'s first-order codes, second-order themes and aggregate dimensions to present our findings (see Table 2).

Table 2. Empirical evidence of factors facilitating ETs on farms

First-order codes	Second-order themes	Aggregation dimensions
"The IoTs are widely used on our farms."	Wide adoption of IoTs	Technology
"We should use this simple-to-operate, fool-proof technical equipment."	Easy-to-use	
"The ETs should have low maintenance rates."	High reliability	
"What kind of model should be used to achieve precise control?"	Internalized agricultural models and software	
"The application of ETs requires technology extension services."	Associated technology extension services	
"We do not have mature management experience in the maintenance and use of equipment."	Management experience of ETs	Organization
"If the farmer has more than 200 acres of farmland, he may recognize these ETs."	Increasing farm size	
"High-quality farmland should be built to support ETs adoption."	Construction of high-quality farmland	

(*continued*)

Table 2. *(continued)*

First-order codes	Second-order themes	Aggregation dimensions
"Our leaders' support is important for us to deploy ETs."	Top management team's support	
"Subsidies are provided to farmers who adopt ETs."	Agricultural equipment subsidies	Environment
"There are policies to support agricultural equipment development."	Increasing R&D for agricultural equipment	
"The local government provides training for farmers."	Training for new farmers	
"The cultural environment also contributes to ET deployment."	Hierarchical cultural environment	
"Governments should widely deploy platforms to disseminate knowledge and information."	Wide deployment of digital platforms to disseminate information	

4 Findings and Discussion

In this study we identify 14 factors facilitating adoption of ETs on farms. For example, from the technological perspective, five factors contribute to their deployment: wide adoption of IoT, easy-of-use, high reliability, internalized agricultural models and software, and associated technology extension services. From the organizational perspective, four impactful factors are identified: management experience of ETs, increasing farm size, top management team's support and construction of high-quality farmland. Finally, from the environmental perspective, we identify agricultural equipment subsidies, increasing R&D for agricultural equipment, training for new farmers, the hierarchical cultural environment, and wide deployment of digital platforms to disseminate information. Our findings differ from those of previous studies. For example, [44]'s investigation suggests that only efficiency and expected competitive advantage determine companies' intentions to use ETs, while [45] reveal six dimensions that may determine ET adoption: ICT infrastructure, human capital, interoperability, big data management, data sharing and data security. Our findings indicate that successful adoption of ETs on farms depends on the technological, organizational and environmental dimensions collectively. Amongst these, the environmental dimension is responsible for providing policy, financial and educational resources to farms and fostering competition between them. The organizational dimension is responsible for developing suitable conditions for ET deployment. Finally, ETs should be reliable and easy-to-use to encourage organizations to deploy them (see Fig. 1).

Our factor identification also extends scholarly understanding. For example, [46] reveal 12 technology adoption enablers on dairy farms, including competitive pressure for adoption, perceptions of the technology's easy-of-use, and availability of a trained workforce. [47] identify 12 enablers of AI adoption and implementation in production

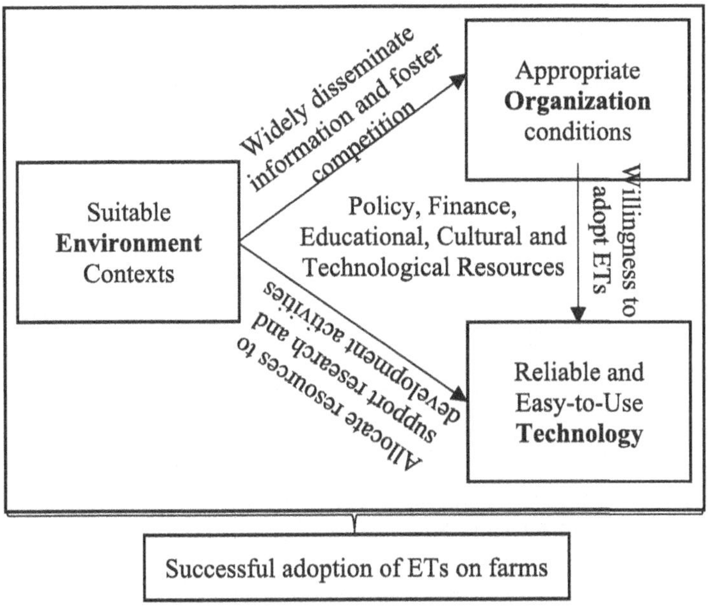

Fig. 1. Evaluated TOE framework

systems, such as external pressure, support from vendors, organizational culture and ethics. Our study differs from previous studies in taking farm size, land quality and external cultural environment into consideration. For example, China's hierarchical cultural value orientation emphasizes fulfilling one's obligations in a hierarchical system and obeying the expectations of those in higher-status roles [48]. In this environment, people view competition as good, which strengthens competition between organizations and forces them to develop appropriate conditions to deploy ETs. Other measures, such as wide deployment of digital platforms (e.g., traceability, information sharing, and production and sales docking platforms) to disseminate information also contribute to successful adoption. One interviewee stated: "*Various digital platforms are built to share technology, product and traceability information. Thus, people using these platforms can receive responses from the market', land' and farmer's response as quickly as possible.*"

5 Conclusions and Research Directions

Grounded in the TOE decision framework, in this study we adopted an inductive, qualitative approach to explore factors affecting ET adoption on farms. We conducted interviews with Chinese farmers and employed thematic analysis to identify factors. Our findings produce novel insights into ET adoption on farms. For example, several factors identified in this study have seldom been mentioned by previous scholars, including internalized agricultural models and software, increasing farm size, construction of high-quality farmland, the hierarchical cultural environment, and wide deployment of digital platforms to disseminate information. We also contribute to the TOE decision framework by proposing that the technological, organizational and environmental dimensions collectively determine successful adoption of ETs on farms. For example, a suitable environment enables wide dissemination of information on ETs, fosters competition between organizations and encourages allocation of resources to support and accelerate ET development. Our findings also have managerial implications for farmers and policymakers. For example, only farmers with more than 200 acres of farmland are likely to invest in and adopt ETs.

Although we employed a rigorous research methodology to explore the phenomenon, our study has limitations. First, it has generalization issues because we collected data only from China. To tackle this limitation, future studies might use questionnaires and evaluate research results from other countries to extend generalizability. Second, in this study we relied on qualitative data collection and analysis methods, which are widely considered to be weaker than a mixed-method approach [49]. Future studies might use multi-criteria decision-making (MCDM) methods to prioritize the factors identified and produce stronger evidence. Third, we conducted semi-structured interviews with 10 Chinese farmers to collect data. However, this sample size may not be sufficient to achieve data saturation, particularly given China's vast agricultural landscape, diverse farming activities, and large farm sizes. Future research could expand the number of interviews to encompass a wider range of farming practices that have adopted ETs, thereby enhancing the robustness and generalizability of the findings.

References

1. Loizou, E., et al.: The role of agriculture as a development tool for a regional economy. Agric. Syst. **173**, 482–490 (2019)
2. Zhang, L et al.: Direct and indirect impacts of urbanization on vegetation growth across the world's cities. Sci. Adv. **8**(27), ebbo0095 (2022)
3. Zhao, G., et al.: Identification, establishment of connection, clustering of social risks involved in the agri-food supply chains: a cross-country comparative analysis. Ann. Oper. Res. **338**, 1241–1281 (2024)
4. Zhao, G., et al.: Risk analysis of the agri-food supply chain: a multi-method approach. Int. J. Prod. Res. **58**(16), 4851–4876 (2020)
5. Springmann, M.: Options for keeping the food system within environmental limits. Nat. **562**, 519–525 (2018)
6. Gupta, D., et al.: Role of existing and emerging technologies in advancing climate-smart agriculture through modeling: a review. Ecol. Inform. **71**, 101805 (2022)
7. Zhao, G., et al.: Blockchain technology in agri-food value chain management: a synthesis of applications, challenges and future research directions. Comput. Ind. **109**, 83–99 (2019)
8. Javaid, M., et al.: Enhancing smart farming through the applications of agriculture 4.0 technologies. Int. J. Intell. Syst. **3**, 150–164 (2022)
9. Giua, C., et al.: Smart farming technologies adoption: which factors play a role in the digital transition? Technol. Soc. **68**, 101869 (2022)
10. Smidt, H.C., Jokonya, O.: Factors affecting digital technology adoption by small-scale farmers in agriculture value chains (AVCs) in South Africa. Inf. Technol. Dev. **28**(3), 558–584 (2022)
11. Giua, C., et al.: Management information system adoption at the farm level: evidence from the literature. Brit. Food. J. **123**(3), 884–909 (2021)

12. Osrof, H.Y., et al.: Adoption of smart farming technologies in field operations: a systematic review and future research agenda. Technol. Soc. **75**, 102400 (2023)
13. John, D., et al.: A systematic review on the factors governing precision agriculture adoption among small-scale farmers. Outlook Agric. **52**(4), 469–485 (2023)
14. Marshall, A., et al.: Critical factors of digital AgTech adoption on Australian farms: from digital to data divide. Inf. Commun. Soc. **25**(6), 868–886 (2022)
15. Dibbern, T., et al.: Main drivers and barriers to the adoption of digital agriculture technologies. Smart Agr. Technol. **8**, 100459 (2024)
16. Giorgio, A.: Enablers to digitalization in agriculture: a case study from Italian field crop farms in the Po River Valley, with insights for policy targeting. Agriculture **14**, 1074 (2024)
17. Ferrari, A., et al.: Drivers, barriers and impacts of digitalization in rural areas from the viewpoint of experts. Inf. Softw. Technol. **145**, 106816 (2022)
18. Sen, L.T.H.: Barriers and enablers of digital extension services' adoption among smallholder farmers: the case of Cambodia, the Philippines and Vietnam. Int. J. Agric. Sustain. **22**(1), 2368351 (2024)
19. Cozzens, S.E., et al.: Emerging technologies: quantitative identification and measurement. Technol. Anal. Strateg. Manage. **22**(3), 361–376 (2010)
20. Stahl, B.C.: What does the future hold? A critical review on emerging information and communication technologies and their social consequences. In: Chiasson, M., Henfridsson, O., Karsten, H., DeGross, J.I. (eds). Researching the Future in Information Systems: IFIP WG 8.2 Working Conference, Future IS 2011. Turku, Finland, June 6–8, 2011
21. Small, H., et al.: Identifying emerging topics in science and technology. Res. Policy **48**(8), 1450–1467 (2014)
22. Rotolo, D., et al.: What is an emerging technology? Res. Policy **44**, 1827–1843 (2015)
23. Morris, W., et al.: Farm diversification, entrepreneurship and technology adoption: analysis of upland farmers in Wales. J. Rural. Stud. **53**, 132–143 (2017)
24. Silveira, F.D., et al.: An overview of agriculture 4.0 development: systematic review of descriptions, technologies, barriers, advantages, and disadvantages. Comput. Electron. Agric. **189**, 106405 (2021)
25. Zhao, G , et al.: Analysis of barriers to adopting Industry 4.0 technologies to achieve agri-food supply chain sustainability: a group-based fuzzy analytic hierarchy process. Bus. Strateg. Environ. **33**(8), 8559–8586 (2024).
26. Romanello, R., Veglio, V. : Industry 4.0 in food processing: drivers, challenges and outcomes. Brit. Food. J. **124**(13), 375–390 (2022)
27. Costa, F., et al.: Industry 4.0 digital technologies enhancing sustainability: applications and barriers from the agricultural industry in an emerging economy. J. Clean. Prod. **408**, 137208 (2023)
28. Kumar, M., et al.: Enablers to achieve zero hunger through IoT and blockchain technology and transform the green food supply chain systems. J. Clean. Prod. **405**, 136894 (2023)
29. Zhao, G., et al.: Understanding the drivers of Industry 4.0 technologies to enhance supply chain sustainability: insights from the agri-food industry. Inf. Syst. Front. (2024). https://doi.org/10.1007/s10796-024-10539-1
30. Kouhizadeh, M., et al.: Blockchain technology and the sustainable supply chain: theoretically exploring adoption barriers. Int. J. Prod. Econ. **231**, 107831 (2021)
31. Galati, A., et al.: Exploring the common blockchain adoption enablers: the case of three Italian wineries. Brit. Food. J. **33**(4), 578–596 (2021)
32. Scur, G., et al.: Analysis of IoT adoption for vegetable crop cultivation: multiple case studies. Technol. Forecast. Soc. Change. **191**, 122452 (2023)
33. Panghal, A., et al.: A study on adoption enablers of 3D printing technology for sustainable food supply chain. Manag. Environ. Qual. **34**(4), 943–961 (2023)

34. Ali, I., et al.: Understanding the key determinants of IoT adoption for the digital transformation of the food and beverage industry. Ind. Manag. Data Syst. **123**(7), 1887–1910 (2023)
35. Kumar, S., et al.: Integrated blockchain and internet of things in the food supply chain: adoption barriers. Technovation **118**, 102589 (2022)
36. Barriball, K.L.: Collecting data using a semi-structured interview: a discussion paper. J. Adv. Nurs. **19**, 328–335 (1994)
37. Zhao, G., et al.: Analysis of factors affecting cross-boundary knowledge mobilization in agri-food supply chains: an integrated approach. J. Bus. Res. **164**, 114006 (2023)
38. Kenny, G.: Customer surveys are no substitute for actually talking to customers. Harv. Bus. Rev. (2019). https://hbr.org/2019/01/customers-surveys-are-no-substitute-for-actually-talking-to-customers
39. O'Keeffe, J., et al.: The use of semi-structured interviews for the characterisation of farmer irrigation practitioners. Hydrol. Earth Syst. Sci. **20**, 1911–1924 (2016)
40. Jayawickrama, U., et al.: Empirical evidence of an integrative knowledge competence framework for ERP systems implementation in UK industries. Comput. Ind. **82**, 205–223 (2016)
41. Nowell, L.S., et al.: Thematic analysis: striving to meet the trustworthiness criteria. Int J Qual Methods **16**, 1–13 (2017)
42. Braun, V., Clarke, V.: Using thematic analysis in psychology. Qual. Res. Psychol. **3**, 77–101 (2006)
43. King, N., Horrocks, C.: Interviews in Qualitative Research. Sage, London (2010)
44. Cater, T et al. (2021). Industry 4.0 technologies usage: motives and enablers. J. Manuf. Technol. Manag. **32**(9), 323–345
45. Adebanjo, D et al. (2023). Key enablers of Industry 4.0 development at firm level: findings from an emerging economy. IEEE. Trans. Eng. Manag. **70**(2), 400–416
46. Kaushik, H., Rajwanshi, R.: Examining the linkages of technology adoption enablers in context of dairy farming using ISM-MICMAC approach. Res. World. Agric. Econ. **4**(4), 68–78 (2023)
47. Merhi, M.I., Harfouche, A.: Enablers of artificial intelligence adoption and implementation in production systems. Int. J. Prod. Res. **62**(15), 5457–5471 (2023)
48. Schwartz, S.H.: A theory of cultural value orientations: explication and applications. Comp. Sociol. **5**(2–3), 137–182 (2006)
49. Creswell, J.W.: Research design, qualitative, quantitative, and mixed methods approaches, 4th edn. Sage Publications, London (2014)
50. Baker, J.: The technology-organization-environment framework. In: Dwivedi, Y., Wade, M., Schneberger, S (eds). Information Systems Theory. Integrated Series in Information Systems. New York: Springer (2011)
51. Yoon, C., et al.: Factors affecting adoption of smart farms: the case of Korea. Comput. Hum. Behav. **108**, 106309 (2020)
52. Sood, A., et al.: Towards sustainable agriculture: key determinants of adopting artificial intelligence in agriculture. J. Decis. Syst. **33**(4), 833–877 (2024)

Exploring the Role of Peer Learning Cycle in Supporting Decision-Making for Food Sharing Organisations

Xinyi Li(✉), Dulekha Kasturiratne, and Huilan Chen

University of Plymouth, Plymouth PL4 8AA, UK
xinyi.li@plymouth.ac.uk

Abstract. Recent global events such as the COVID-19 pandemic, the political conflict, and ongoing living crisis and inflation have increasingly driven individuals and families into the critical state of food insecurity. To rescue the situation, food sharing initiatives such as food sharing organisations and projects have been implemented across many countries. One of the primary challenges faced by food sharing organisations is the lack of decision-making support to enhance operational efficiency and create community value. In order to address this issue and provide a well-grounded theoretical approach for support decision-making and create community value in food sharing communities, this paper investigated the role of peer learning cycle in supporting decision making for food sharing organisations and creating closed-loop community networks in food sharing communities. A Systematic Literature and Practice Review (SLPR) has been undertaken. Based on the SLPR, the PeersForFood framework has been developed to illustrate how a peer learning cycle model can support decision-making processes in food sharing organisations and facilitate the creation of closed-loop community value networks within food sharing communities. The PeersForFood framework contributes to new knowledge and practice, offering valuable insights and efficient recommendations for decision-makers in food sharing organisations and strengthening the resilience of food sharing communities.

Keywords: Decision Support · Peer Learning · Food-sharing Organisations · Food-sharing Communities · Community Value Network · Systematic Literature and Practice Review

1 Introduction

Hunger represents both a profound violation of human dignity and a significant obstacle to social, political, and economic advancement (FAO Council, 2010). According to the United Nations World Food Programme (2023), recent global events such as the COVID-19 pandemic, the political conflict, as well as the ongoing cost-of-living crisis

Note: not all 72 papers included in the SLPR but only selected key references are listed here because of page limit.

B. Delibašić et al. (Eds.): ICDSST 2025, LNBIP 546, pp. 15–28, 2025.
https://doi.org/10.1007/978-3-031-90863-7_2

and inflation, have increasingly driven individuals and families into the critical state of food insecurity (FAO. et al., 2024). In order to address food insecurity, many countries have implemented various food sharing initiatives (Mackenzie & Davies, 2019; Lombardi & Costantino, 2020). These include the establishment of food sharing organisations and projects (e.g. CULTIVATE, FoodSHIFT2030, SHARECITY), alongside the introduction of ICT-mediated food sharing online platforms and apps for public use. Such initiatives have demonstrated success in enhancing food crisis support and fostering food sustainability and democracy (Jaeggi & Gurven, 2013, Mackenzie & Davies, 2019). Existing studies on food sharing have primarily focused on its impact on reducing food waste and promoting food sustainability (Davies et al., 2019; Mackenzie & Davies, 2019; Zhao et al., 2023), as well as the role of ICT-mediated technologies in facilitating food sharing activities (Harvey et al., 2020; Mazzucchelli et al., 2021).

One of the primary challenges faced by food sharing organisations is the lack of decision-making support to enhance operational efficiency and create community value. Unlike long-established and professional organisation types, such as business, educational, or governmental organisations, the research field on food sharing organisations are relatively nascent, thus there is limited empirical evidence to support the decision-making and operational efficiency in food sharing organisations since information sharing strategies is underdeveloped, which eventually may prohibit the creation of social value at the community level (Lombardi & Costantino, 2020; Morone et al. 2018). Thus, there is an urgent need for research aimed at exploring solutions to enhance operational efficiency and decision making in food sharing organisations, also create community value for food sharing communities while strengthening their capacity and resilience to meet the growing demand for sustainability (Ansari et al., 2012; Berns et al., 2023; Davies et al., 2018; Davies et al., 2019). To address this gap, this study aims to investigate the role of peer learning cycle in supporting decision making for food sharing organisations and creating closed-loop community networks in food sharing communities. Through a systematic literature and practice review on current literature and practical initiatives, the objective is to identify specific components of the peer learning cycle and the closed-loop community value network. The PeersForFood framework offers valuable insights and efficient recommendations for decision-makers in food sharing organisations and strengthens the capacity and resilience of food sharing communities.

2 Methodology: Systematic Literature and Practice Review

A Systematic Literature review (SLR) was considered the most suitable method for this paper due to its comprehensive and structured nature. Unlike traditional literature reviews, Systematic literature review (SLR) is a rigorous approach that allows researchers to synthesize and refine scattered knowledge from established literature and generate new insights and theoretical contributions (Tranfield et al., 2003). Given the practical nature of this research, this paper also reviews practical initiatives, such as previous and ongoing recognised food sharing projects and food sharing organisations. These include project reports (e.g. CULTIVATE), publications (e.g. FoodSHIFT2030), project databases (e.g. SHARECITY100), and official websites (e.g. Feeding Devon, Feeding Britain, FareShare, etc.). This systematic review combination of academic literature and practical sources (hence SLPR) provides a comprehensive and evidence-based foundation for the analysis of peer learning and its influence on decision making and community value creation. The SLPR process is divided into 5 phases and each phase will be described in detail as follows (see Fig. 1):

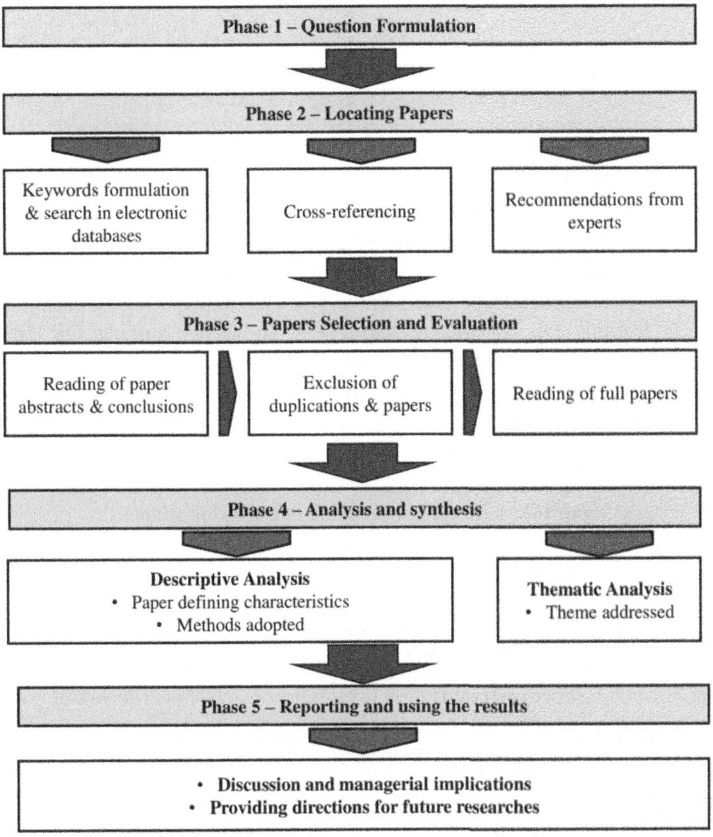

Fig. 1. SLPR methodology adopted in the study

The first step of a SLPR is to formulate precise, informative and clearly defined research questions to ensure clarity and avoid ambiguity (Hohenstein et al., 2015):

- *RQ1.* What are the main components of peer learning and the community value creation network?
- *RQ2.* How does peer learning facilitate decision making in food sharing organisations, ultimately impact the creation of community value in food sharing communities?

The second step involves locating relevant high-quality academic papers to form a comprehensive list of key contributions aligned with the research questions (Denyer &

Tranfield, 2009). Trusted academic databases, including Web of Science, Scopus, ScienceDirect, Taylor & Francis, and Wiley Online Library, were selected for their world-leading reputation and relevance in business research. Several defined keywords (e.g. "peer learning", "food sharing", and "community value", with all the related terms) were used as the search criteria. These terms were searched in the title, abstract, and keywords (see Table 1).

Table 1. Databases and search strings

Databases	Web of Science, Scopus, ScienceDirect, Taylor & Francis, Wiley Online Library, Cordis
Search strings	('food sharing' OR 'food waste' OR 'food redistribution' OR 'food democracy' OR 'food donation' OR 'food sustainability' OR 'food pantry' OR 'food bank' OR 'sharing economy' OR 'urban food' OR 'free food') AND ('peer learning' OR 'peer mentor' OR 'peer to peer study' OR 'community learning' OR 'knowledge sharing' OR 'knowledge learning' OR 'knowledge transfer') AND ('community value' OR 'social value' OR 'societal value' OR 'social innovation' OR 'value creation' OR 'value generate' OR 'create value' OR 'value network')

Table 2. Criteria for inclusion or exclusion studies

Criteria	Inclusion	Exclusion
	Identification Stage	
Availability	Full-text available	No full-text available
Peer-review	Peer-reviewed	Not peer-reviewed
Type of publication	Journal articles, conference papers and book chapters	Reviewed articles, posters, etc.
Year of publication	2005 - 2025	Before 2005
Language	English Language	Non-English
	Screening stage	
Types of study	Theoretical and conceptual studies	Discussion papers, overviews, opinion papers, etc.
Relevance	Studies relevant to research questions (e.g. contain keywords, same/similar research focus, etc.)	Studies that are not relevant to research questions

After identifying papers using the search strings from the chosen databases, a series of explicit inclusion and exclusion criteria were applied to paper selection and evaluation in the subsequent phase (see Table 2). These criteria considered elements such as year and type of publication, research relevance and overall paper quality. This stringent evaluation process ensured the selected studies that were both highly relevant to the research questions and methodologically robust, thereby enhancing the overall reliability of the findings. Studies that do not meet the inclusion criteria or duplicates were eliminated. Then, the papers were screened by title, abstract, introduction, and then based on all remaining articles. After adding additional relevant papers by cross-referencing all the citations and references, a total of 72 academic peer-reviewed journal papers, conference papers, and book chapters to be included in the analysis. The overall literature search and selection process is illustrated in Fig. 2.

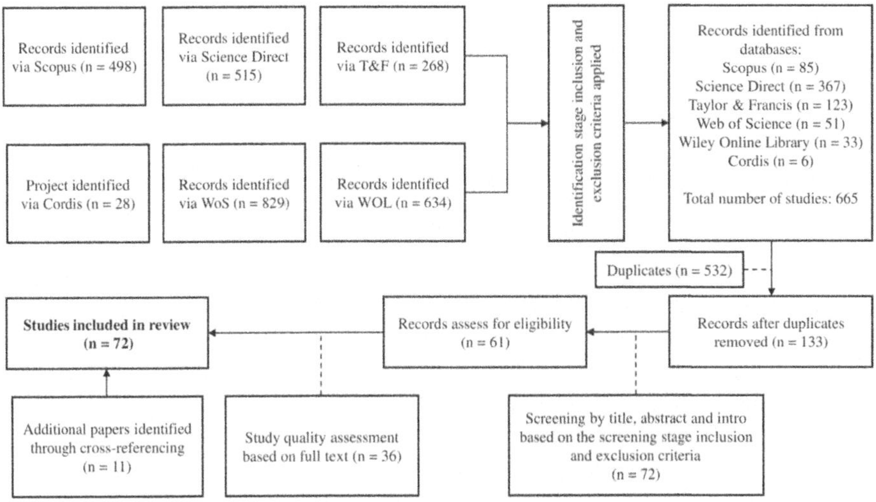

Fig. 2. Paper Selection flowchart

The next stage encompasses of data extraction and finding synthesis, key information was extracted from the selected papers in order to answer research questions, identify recurring themes and notable gaps in the existing literature. This process offered a holistic understanding of the role of peer learning cycle in supporting decision-making and operational efficiency in food sharing organisations and contributes to the development of closed-loop community value networks within food sharing communities. Building on the comprehensive insights gained from the SLPR, the subsequent section presents key findings from the current literature and practices, then developing a conceptual framework that designed to facilitate decision making in food sharing organisations, which ultimately impact the creation of community value in food sharing communities.

3 Key Findings

Based on the SLPR results, this section is divided into three key themes: food sharing and its inner system, organisations and communities, the application of peer learning in food sharing communities, and the concept of community value networks in food sharing communities.

3.1 Food Sharing: The System, Organisation and Community

The food sharing project SHARECITY, conducted from 2016 to 2021, identified various typologies of food sharing. These include not only the exchange of food products but also the facilitation of connections among individuals interested in sharing land and tools for food cultivation. Additionally, it encompasses the sharing of kitchen spaces, food preparation, and storage equipment. Importantly, food sharing facilitates the exchange of knowledge, skills, and information related to food availability, as well as techniques for growing, processing, and cooking food (Davies, 2016; Davies et al., 2018). Consequently, food sharing is often perceived as a collective action against issues such as food waste, food crisis and hunger, and the formation of food sharing systems is viewed as an activist response to mitigate the consequences of capitalist and global food chains (Berns et al., 2023; Schanes & Stagl, 2019).

The food sharing system comprises three main stakeholders: food donors on the supply side (e.g., supermarkets, restaurants, food retailers, etc.), food sharing organisations as the connector (e.g., food banks, local charities, kitchen hubs, etc.), also food receivers on the demand side (e.g., population in need, local communities, schools, etc.). As the joint point of the food sharing system, food sharing organisation is the focal point of the current analysis. Food sharing organisations play a critical role in facilitating the distribution of food between donors and receivers, thereby increasing access to food and creating long-lasting networks of support at both local and national levels. For example, food sharing organisations like *Feeding Devon*, *Feeding Britain*, and *FareShare* act as connectors within the food sharing systems, ensuring that excess food from donors reaches those who need it most. They help alleviate food insecurity and reduce food waste by redistributing surplus food. Research suggests that food sharing organisations not only bridge the gap between food supply and demand but also strengthen community ties by fostering a culture of sharing, which can improve the resilience and sustainability of local food systems (Davies, 2016; Davies et al., 2018). These organizations contribute to more equitable food distribution and play a vital role in addressing hunger and food insecurity, particularly in disadvantaged communities.

Local food sharing communities, often formed through activities organised by food sharing organisations, are integral to addressing both food access and social connectivity. These communities are frequently centered around shared goals of healthy eating and lifestyles, and they serve as platforms for mutual support and collaboration (Ganglbauer et al., 2014; Parker et al., 2010). In areas where poverty and social disadvantage are prevalent, such as rural or low-income neighbourhoods, food sharing communities often include disadvantaged groups such as households with low-income, one-parent families, refugees, and the homeless (FAO et al., 2024). For example, Feeding Devon (2024) reports that many users of their community kitchen and food bank services are low-income families dealing with financial instability, lone parents experiencing benefit delays, and newly housed homeless individuals and families. Thus, local food sharing communities not only provide its members with critical access to food but also serve as "social networking sites," where people with shared values can meet, interact, and build connections (Ganglbauer et al., 2014; Lozano, 2007). In this case, local food sharing communities contribute to both the practical aspects of food crisis support and the social fabric of the communities they serve, offering both nourishment and a sense of belonging and support.

3.2 Peer Learning in Food Sharing Communities

Peer learning has been widely advocated as an effective approach across a variety of sectors such as education, health, and finance, with scholars and policymakers recognising it as an effective method for knowledge exchange and information sharing (Donald & Ford, 2023; Oliver et al., 2023; Saetova et al., 2018; Song et al., 2020). Unlike traditional learning methods that are typically confined to specific contexts like classrooms or schools, peer learning can happen spontaneously in everyday life, unrestricted by location, participants, or skill levels. Existing literature identifies four key components of peer learning: agents, modes, contents, and outcomes.

Peer learning agents can be divided into two categories: human agents and boundary objects. Human agents are typically experienced individuals who bring knowledge and expertise to the learning process (Oliver et al., 2023; Petosa & Smith, 2014), while boundary objects, refers to structured resources that can support the peer learning process and facilitate the knowledge exchange process across boundaries, such as handbooks, videos, and toolkits (Rigg et al., 2021). These agents are essential in facilitating the transfer of knowledge and skills. The modes of peer learning encompass various approaches, from organisation-based learning, such as peer presentations and training workshops, to cross-organisation learning, including conferences, solution forums, and ICT-mediated discussion platforms (Matsuo & Aihara, 2022; Rigg et al., 2021; Saetova et al., 2018; Topping, 2005; Zhao et al., 2023). These modes ensure the broad applicability of peer learning, reaching different groups across various settings.

The content of peer learning is highly context-dependent, often tailored to the specific needs of participants. However, it commonly revolves around knowledge sharing and the development of skills in particular areas, making it especially relevant for communities seeking to enhance practical expertise. The outcomes of peer learning are evaluated at both individual and organisational levels. On an individual level, peer learning fosters personal growth, skill development, and the building of connections within networks. At the organisational level, it contributes to enhanced operational efficiency and decision-making, promotes collaborative innovation, and strengthens organisational culture and engagement (Donald & Ford, 2023; Oliver et al., 2023; Rigg et al., 2021; Topping, 2005). These outcomes generate a feedback loop, where the experiences of participants contribute to a cyclical process of improvement. This continuous exchange not only benefits the learners but also the organisations they belong to, facilitating an ongoing process of development that drives collective progress.

Food sharing organisations play a pivotal role in the food sharing system, functioning as "community hubs" where food sharing activities are often coordinated (Berns et al., 2021; Davies et al., 2019). Local food sharing organisations serve as focal points for individuals to engage in shared practices and participate in communal activities, fostering not only social relationships but also learning through these interactions. As Lave and Wenger (1991) and Fox (2000) note, learning is inherently social and can be nurtured through active involvement in community practices. Within the context of food sharing, peer learning can play an integral role in building networks, enhancing skills, and disseminating knowledge across communities.

Thus, peer learning is particularly relevant in food sharing communities, as it offers a multi-way, mutually beneficial approach for knowledge exchange and collaborative decision making. It enables the removal of knowledge boundaries, facilitates the co-creation of community values, and accelerates the sharing and acquisition of knowledge (Liu, 2020). By participating in food sharing activities, individuals not only gain practical skills but also contribute to the broader community value by sharing experiences and insights, thus creating a dynamic environment for continuous learning and growth.

3.3 Community Value Creation, Measurement, Capture, and Feedback in Food Sharing Communities

Community value, as a specific form of social value, arises from collective activities and interactions within communities and plays a significant role in sustaining community networks and well-being (Mair & Martí, 2006; Pret & Carter, 2017; Santos, 2012; Shaw & Carter, 2007). Research has demonstrated that community value significantly contributes to social value creation in multiple ways, including fostering community cohesion, community engagement and empowerment (Santos, 2012; Wenger et al., 2011). Community cohesion is closely tied to the development of human and social capital, evidenced by the establishment of social relationships, networks, and the creation of positive spillover effects within the community (Hietschold et al., 2023). Community engagement focuses on fostering inclusive participation, collaborative efforts, and active involvement among community members. This dynamic participation supports the equitable distribution of resources and the development of inclusive practices that enable communities to thrive collectively (Morrow, 2019; Mackenzie & Davies, 2019; Rut et al., 2021). Meanwhile, community empowerment involves both the recognition of opportunities and the empowerment and acknowledgment of community members, which strengthens their sense of identity and involvement (Berns et al., 2023; Donald & Ford, 2023; Rut et al., 2021). These dimensions of community value underline its integral role in broader social value creation, highlighting how active participation in community activities not only strengthens individual and collective well-being but also enhances societal resilience and sustainability.

Existing literature has identified three specific aspects of social value creation within communities: address community needs, developing community capabilities, and measure its outcome and feedback (Ansari et al., 2012; Dembek et al., 2016; Lashitew et al., 2022). In the context of food sharing communities, these dimensions translate into a series of interconnected processes. First, it is essential to articulate the primary value propositions and identify the beneficiaries of these values. Next, the focus should shift to determining and fostering the necessary capabilities within the community to facilitate the creation of value. Finally, mechanisms for capturing and measuring the outcomes must be established to ensure the reintegration of these benefits into the community. By prioritising the creation, capture, and reinvestment of community value through a closed-loop approach, food sharing communities can bolster their resilience, enhance sustainability, and generate long-lasting benefits.

4 The 'PeersForFood' Conceptual Framework

Drawing on the existing literature review, the 'PeersForFood' conceptual framework has been developed to illustrate the role of a peer learning cycle model in supporting decision making in food sharing organisations and the community value network within food sharing communities (as shown in Fig. 3). In the conceptual framework, the four key sub-sections in the peer learning section are based on the circulated peer learning operation cycle model mentioned earlier in the key findings: peer learning agents (Oliver et al., 2023; Petosa & Smith, 2014), peer learning modes (Rigg et al., 2021; Saetova et al., 2018), peer learning capabilities and peer learning outcomes (Donald & Ford, 2023; Topping, 2005). Each component encompasses multiple elements based on varying dimensions (e.g., individual vs organisational levels; organisation-based vs cross-organisation-based approaches) and types (e.g., human participants and boundary objects). Thus, the PeersForFood framework offers comprehensive theoretical evidence and potential solutions for decision-makers in food sharing organisations to foster operational efficiency and better decision making.

Within the peer learning cycle model, certain elements have already been implemented in practice. For instance, food sharing organisations like *Feeding Devon* have introduced experienced individuals as 'advice workers' to their newly established food hubs, while Feeding Britain has utilised 'Feeding Britain toolkit' as boundary objects to create and develop their Affordable Food Clubs. Peer learning modes, such as training workshops and conferences, are widely adopted across food sharing organisations, facilitating the exchange of knowledge and best practices, particularly in areas like food sharing activities and organisational management (Rigg et al., 2021). In contrast to these well-documented elements and components, peer learning capabilities represent a relatively novel concept. These capabilities focus on identifying and enhancing skills specific to food sharing organisations, such as information-sharing, communication, and decision-making capabilities. Through the utilisation of peer learning agents and modes, these capabilities are expected to grow and evolve, enabling organisations to innovate and acquire new capabilities. As a result, peer learning outcomes are anticipated to manifest at both individual and organisational levels. For individuals, this includes personal growth, knowledge and skill development, and the establishment of connections and networks (Donald & Ford, 2023; Oliver et al., 2023). For organisations, previous studies have shown that peer learning can enhance operational efficiency, foster collaborative decision making, and strengthen organisational culture and engagement (Rigg et al., 2021; Topping, 2005). Collectively, these components create a cohesive cycle model designed to support and sustain community value networks within food sharing communities.

The community value (i.e. cv) network section in the conceptual framework consists of four components, including define CV proposition, map CV beneficiaries, CV creation and delivery, and CV measure and feedback, each with detailed elements (Morrow, 2019; Rut et al., 2021; Santos, 2012; Wenger et al., 2011). For example, in order to define community value propositions, it is essential to address the needs of food sharing communities. For example, The UN World Food Programme (2023) identified some crucial needs for food sharing communities, such as improving food crisis support, fostering food sustainability, and promoting healthy lifestyle. The next step is to map

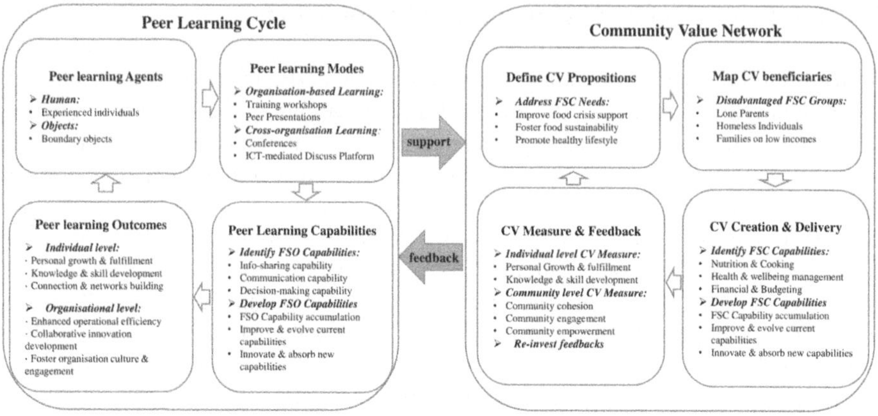

Fig. 3. PeersForFood Conceptual Framework

community value beneficiaries and identify disadvantaged groups within the food sharing community. According to Feeding Devon (2024), there are three disadvantaged groups that have highly participated in food sharing activities and communities, which are lone parents, homeless individuals and families on low incomes.

Then, community values are expected to be created and delivered through identifying and developing capabilities of food sharing communities. There are three main capabilities that have been identified in food sharing communities: nutrition and cooking capability, health and wellbeing management capability, financial and budgeting capability (Feeding Devon, 2004; Lashitew et al., 2022). Created community value can be measured in both individual and community level, using criteria such as persona growth and fulfilment, knowledge and skill development, community cohesion, community engagement and community empowerment (Berns et al., 2023; Hietschold et al., 2023; Rut et al., 2021). To close the loop, all the measure results and will feedback to food sharing communities. Again, both these aspects and components are selected based on the existing literature and actual practice, working as preliminary indicators in this paper. Overall, this conceptual framework providing a comprehensive understanding of the application of peer learning cycle and how it will support decision making in food sharing organisations and impact the creation of a closed-loop community value network within food sharing communities.

5 Discussion: Research Gaps and Future Research Directions

This section aims to address several key areas that remain underexplored, offering valuable insights for advancing research and contributing to the development of existing knowledge in this field.

Regarding the role of peer learning in food sharing organisation, potential mechanisms were identified and discussed in the key findings section. While peer learning is widely recognised as an effective approach across various domains, its implementation in the context of food sharing organisations remains insufficiently examined. Although some studies have acknowledged the potential of peer learning to enhance food sharing practices and sustainability (Meshulam et al., 2023; Morrow, 2019; Oliver et al., 2023), there is limited research investigating the role of peer learning in improving capability development as a key outcome. Furthermore, the introduction of a peer learning cycle model within the context of food sharing organisations has not been adequately addressed, which provides a valuable direction for further exploration. For example, real-world testing could be conducted through future empirical studies within food sharing organisations to assess the practical applicability and provide empirical evidence of the PeersForFood conceptual framework. By applying the framework in practice, organisations can assess how peer learning enhances operational efficiency, collaboration, and social value creation. Real-world validation also helps identify challenges and refine the framework to better suit organisational needs. Ultimately, this process ensures that the framework is not only theoretically robust but also practically applicable in fostering sustainable decision-making processes and strengthening community value networks, which could yield valuable insights for decision-makers and leaders of food sharing organisations and policymakers.

Furthermore, challenges remain in raising researchers' concern in community value creation in food sharing communities. Previous studies have mainly investigated social innovation and social networking within food sharing communities (Ganglbauer et al., 2014; Harvey et al., 2020; Lombardi & Costantino, 2020), as well as issues related to social inclusion in urban food innovations (Lepiller & Valette, 2021). Nevertheless, there remains a notable gap in the literature focusing on the social value perspective, represents a promising avenue for further research. Future research could put a broader emphasis on solutions for creating community value within food sharing communities, and examine the concept of closed-loop, circular networks that encompass community value creation, capture, and reinvestment processes. Such an approach would provide a more holistic understanding of sustainable development within food sharing communities.

6 Conclusion

In this paper, a SLPR has been conducted to reviewing existing literature and practices in peer learning and community value creation in food sharing organisations and communities. Through a set of process of formulating questions, identifying, selecting and evaluating papers, 72 publications were identified and reviewed in this paper. By synthesising key findings from the existing literature and practices, a PeersForFood conceptual framework with effective elements has been proposed to illustrate the role of peer learning cycle in facilitate decision making in food sharing organisation and create community value in food sharing communities.

Finally, this paper makes theoretical contribution to the current knowledge by proposing a holistic theoretical and practical framework on the implementation of the peer learning cycle in food sharing organisation, which not only provides a useful tool for decision making and capability development in food sharing organisations but also support the creation of community value in the food sharing community. Moreover, this research offers valuable insights for policymakers and food sharing organisations' leaders, informing important recommendations for future actions, ultimately strengthening the food sharing community and its objectives.

Acknowledgements. The work reported in this paper was conducted as part of a University Research Studentship (URS) funded by the University of Plymouth.

References

Ansari, S., Munir, K., Gregg, T.: Impact at the 'bottom of the pyramid': the role of social capital in capability development and community empowerment. J. Manage. Stud. **49**(4), 813–842 (2012)

Berns, K., Rossitto, C., Tholander, J.: Queuing for waste: Sociotechnical interactions within a food sharing community. In: CHI'21 Making Waves, Combining Strengths, 10. Online Virtual Conference (originally Yokohama, Japan): Association for Computing Machinery (2021)

Berns, K., Rossitto, C., Tholander, J.: Learning from other communities: organising collective action in a grassroots food-sharing initiative. Comput. Support. Coop. Work **32**(4), 951–999 (2023)

Davies, A.: SHARECITY Sustainability of city-based food sharing: typologies of food sharing. In: ShareCity Working Paper No. 1. (2016)

Davies, A., Weymes, M., Mackenzie, S.: Communicating goals and impacts of food-sharing. Urban Agric. Mag. **34**, 38–40 (2018)

Davies, A.R., Cretella, A., Franck, V.: Food sharing initiatives and food democracy: practice and policy in three European cities. Polit. Governance **7**(4), 8–20 (2019)

Dembek, K., Singh, P., Bhakoo, V.: Literature review of shared value: a theoretical concept or a management buzzword? J. Bus. Ethics **137**(2), 231–267 (2016)

Denyer, D., Tranfield, D.: Producing a systematic review. In: Buchanan, D., Bryman, A. (eds.) The Sage Handbook of Organizational Research Methods, Chapter 39, pp. 671–689. Sage Publications Ltd, London (2009)

Donald, W.E., Ford, N.: Fostering social mobility and employability: the case for peer learning. Teach. High. Educ. **28**(3), 672–678 (2023)

FAO Council: The Right to Food (2010)

FAO, IFAD, UNICEF, WFP & WHO: The state of food security and nutrition in the World 2024: Financing to end hunger, food insecurity and malnutrition in all its forms. Rome (2024)

Devon, F.: Health Eating Project (2024)

Fox, S.: Communities of practice, foucault and actor-network theory. J. Manage. Stud. **37**(6), 853–868 (2000)

Ganglbauer, E., Fitzpatrick, G., Subasi, Ö., Güldenpfennig, F.: Think globally, act locally: a case study of a free food sharing community and social networking. In: Proceedings of the 17th ACM Conference on Computer Supported Cooperative Work & Social Computing, CSCW'14: Computer Supported Cooperative Work, pp. 911–921 (2014)

Harvey, J., Smith, A., Goulding, J., Branco Illodo, I.: Food sharing, redistribution, and waste reduction via mobile applications: A social network analysis. Ind. Mark. Manage. **88**, 437–448 (2020)

Hietschold, N., Voegtlin, C., Scherer, A.G., Gehman, J.: Pathways to social value and social change: An integrative review of the social entrepreneurship literature. Int. J. Manag. Rev. **25**(3), 564–586 (2023)

Hohenstein, N., Feisel, E., Hartmann, E., Giunipero, L.: Research on the phenomenon of supply chain resilience: a systematic review and paths for further investigation. Int. J. Phys. Distrib. Logistics Manage. **45**(1/2), 90–117 (2015)

Jaeggi, A.V., Gurven, M.: Natural co-operators: food sharing in humans and other primates. Evol. Anthropol. Issues News Rev. **22**(4), 186–195 (2013)

Lashitew, A.A., Narayan, S., Rosca, E., Bals, L.: Creating social value for the 'base of the pyramid': an integrative review and research agenda. J. Bus. Ethics **178**(2), 445–466 (2022)
Lave, J., Wenger, E.: Situated Learning: Legitimate Peripheral Participation. Cambridge University Press, Cambridge, MA (1991)
Lepiller, O., Valette, É.: Urban food innovations – scaling and social inclusion issues. UNESCO Chair in World Food Systems (2021)
Liu, S.: Knowledge Management: An Interdisciplinary Approach for Business Decisions. Kogan Page (2020)
Lombardi, M., Costantino, M.: A social innovation model for reducing food waste: the case study of an Italian non-profit organization. Adm. Sci. **10**, 45 (2020)
Lozano, R.: Collaboration as a pathway for sustainability. Sustain. Dev. **15**(6), 370–381 (2007)
Mackenzie, S.G., Davies, A.R.: SHARE IT: co-designing a sustainability impact assessment framework for urban food sharing initiatives. Environ. Impact Assess. Rev. **79**, 106300 (2019)
Matsuo, M., Aihara, M.: Effect of a community of practice on knowledge sharing across boundaries: the mediating role of learning goals. J. Knowl. Manag. **26**(1), 1–16 (2022)
Mazzucchelli, A., Gurioli, M., Graziano, D., Quacquarelli, B., Aouina-Mejri, C.: How to fight against food waste in the digital era: key factors for a successful food sharing platform. J. Bus. Res. **124**, 47–58 (2021)
Mair, J., Martí, I.: Social entrepreneurship research: a source of explanation, prediction, and delight. J. World Bus. **41**(1), 36–44 (2006)
Meshulam, T., Font-Vivanco, D., Blass, V., Makov, T.: Sharing economy rebound: The case of peer-to-peer sharing of food waste. J. Ind. Ecol. **27**(3), 882–895 (2023)
Morone, P., Pasquale, M.F., Enrica, I., Morone, A.: Does food sharing lead to food waste reduction? An experimental analysis to assess challenges and opportunities of a new consumption model. J. Clean. Prod. **185**, 749–60 (2018)
Morrow, O.: Community self-organizing and the urban food commons in Berlin and New York. Sustainability **11**(13), 36–41 (2019)
Oliver, T.L., Diewald, L.K., McKeever, A., George, C.A., Shenkman, R.: Empowering community leadership: perspectives of peer mentors facilitating a food pantry-based nutrition education program. Int. J. Environ. Res. Public Health **20**(3), 2604 (2023)
Parker, A., Landry, B.M., Grinter, R.E.: Characteristics of shared health reflections in a local community. In: Proceedings of CSCW '10, pp. 435–444. ACM, New York, NY, USA (2010)
Petosa, R.L., Smith, L.H.: Peer mentoring for health behavior change: a systematic review. Am. J. Health Educ. **45**(6), 351–357 (2014)
Pret, T., Carter, S.: The importance of 'fitting in': collaboration and social value creation in response to community norms and expectations. Entrep. Reg. Dev. **29**(7–8), 639–667 (2017)
Rigg, C., Coughlan, P., O'Leary, D., Coghlan, D.: A practice perspective on knowledge, learning and innovation - insights from an EU network of small food producers. Entrep. Reg. Dev. **33**(7–8), 621–640 (2021)
Rut, M., Davies, A.R., Ng, H.: Participating in food waste transitions: exploring surplus food redistribution in Singapore through the ecologies of participation framework. J. Environ. Planning Policy Manage. **23**(1), 34–47 (2021)
Saetova, L., Uulu, Z.Y., Kisunko, G.: Applying community of practice peer-to-peer learning approach in Rural Kyrgyzstan. In: Governance Global Practice Europe and Central Asia Region World Bank, pp. 1–7 (2018)
Santos, F.M.: A Positive Theory of Social Entrepreneurship. J. Bus. Ethics **111**(3), 335–351 (2012)
Schanes, K., Stagl, S.: Food waste fighters: what motivates people to engage in food sharing? J. Clean. Prod. **211**, 1491–501 (2019)
Shaw, E., Carter, S.: Social entrepreneurship: theoretical antecedents and empirical analysis of entrepreneurial processes and outcomes. J. Small Bus. Enterp. Dev. **14**(3), 418–434 (2007)

Song, H.J., Harris, M., Li, F.Y., Green, J., Knight, A.: Connecting general practitioners through a peer-facilitated community of practice for chronic disease care. Ann. Fam. Med. **18**(2), 179 (2020)

The United Nations World Food Programme: Global report on food crises: Number of people facing acute food insecurity rose to 258 million in 58 countries in 2022 (2023)

Topping, K.J.: Trends in peer learning. Educ. Psychol. **25**(6), 631–645 (2005)

Tranfield, D., Denyer, D., Smart, P.: Towards a methodology for developing evidence-informed management knowledge by means of systematic review. Br. J. Manag. **14**(3), 207–222 (2003)

Wenger, E., Trayner, B., Laat, D.: Promoting and Assessing Value Creation in Communities and Networks: A Conceptual Framework. Open Universiteit, Ruud de Moor Centrum, Netherlands, Ruud de Moor Centrum Raporten (2011)

Zhao, G., Liu, S., Wang, Y., Lopez, C., Ong, A., Chen, X.: Reducing food waste from social innovation perspective: a review of measures, research gaps and future directions. Int. Food Agribusiness Manage. Rev. **26**(2), 199–223 (2023)

Decision Support in the Context of Brazilian Plan for Control of Waste and Contaminants in Dairy Industry

Edeneide Laura de Melo Santos[1], Ana Paula Henriques de Gusmão[1,2(✉)], Jônatas Araújo de Almeida[1,3], and Ana Paula Cabral Seixas Costa[4]

[1] PPGEP-CAA, Universidade Federal de Pernambuco, Caruaru, Brazil
{edeneide.laura,jonatas.almeida}@ufpe.br, anapaulahg@academico.ufs.br

[2] Management Engineering Department, Universidade Federal de Sergipe, São Cristóvão, Brazil

[3] MAPS Group - Modelling and Alignment of Portfolio and Strategy Group, Universidade Federal de Pernambuco, Recife, Brazil

[4] Management Engineering Department, CDSID - Center for Decision Systems and Information Development, Universidade Federal de Pernambuco, Recife, Brazil
apcabral@cdsid.org.br

Abstract. The increase in foodborne diseases has sparked interest in food safety. The Brazilian Ministry of Agriculture, Livestock and Food Supply created the National Plan for Residue and Contaminant Control (PNCRC) to assess and minimize the risks of contaminants in food production. This plan includes a formal methodology for defining visits to milk and dairy companies. However, there is no formal guidance for defining the suppliers to be included in the PNCRC. Therefore, this article proposes and applies a multicriteria decision support model to define the suppliers to be inspected. The FITradeoff multicriteria method is incorporated into the model due to its flexible and interactive process to assist in this decision-making process. Finally, a case study was conducted, demonstrating the robustness of the FITradeoff method as an effective tool to support decision-making in the context of the PNCRC.

Keywords: National Plan for Control of Residues and Contaminants · Milk and dairy products · Inspection process · FITradeoff method

1 Introduction

According to the World Health Organization [1], an estimated 600 million people worldwide fall ill each year after eating contaminated food, and more than half of these people die. This organization warns of the impacts in terms of lost productivity and medical expenses, especially in low- and middle-income countries, which further harms the economic development of these countries. Therefore, the need for access to enough safe and nutritious food, combined with population growth, has made food safety one of the biggest global concerns [2, 3].

B. Delibašić et al. (Eds.): ICDSST 2025, LNBIP 546, pp. 29–42, 2025.
https://doi.org/10.1007/978-3-031-90863-7_3

The growing number of foodborne diseases (FBDs) worldwide has increased public interest in food safety, especially when compared to the degree of severity associated with them [4]. Seeking to ensure food safety and quality, the Brazilian Ministry of Agriculture, Livestock and Food Supply (MAPA), through the Federal Inspection Service (SIF), implemented the National Plan for Residue and Contaminant Control (PNCRC). The main objective of this plan is to enable the systematic monitoring of chemical residues that represent a concern for public health in animals intended for slaughter, as well as in milk, eggs, honey and fish processed in establishments registered with the Department of Inspection of Products of Animal Origin (DIPOA) and un-der federal inspection [5].

MAPA determines the frequency of inspections of establishments based on a risk assessment that considers several factors. In Brazil, establishments producing milk and dairy products are the most inspected sector, representing 38% of establishments under federal inspection, followed by the meat sector, which represents 24.9% of inspections [6]. Thus, strict monitoring of the production of milk and dairy products is extremely important for society.

Due to the impossibility of carrying out inspections of all establishments constantly, the inspection method adopted by SIF is based on the risk that the establishment poses due to both its production process and the products it produces, considering the specific behavior of each establishment in terms of compliance with legislation. Based on these aspects, SIF seeks to balance the effectiveness of the actions and the costs spent in this process, optimizing the work carried out [7].

Within the scope of the visits, or according to the sampling plan, some milk producers are defined to also be visited by SIF, when they are included in the sampling defined by the PNCRC. Although there is a formal methodology for defining visits to producing companies, there is no formal methodology for defining the suppliers to be included in the PNCRC. The criteria that are used to define the suppliers that will be visited vary depending on the SIF team and other circumstances, although the objective is always to minimize the risk of contaminants.

In the literature, it is possible to find some studies that contribute to risk assessment, in the context of food safety, and therefore direct inspections by the responsible bodies [8, 9]. However, as reported in [10], the inspection methods adopted by food security authorities in different countries have shouting differences. In the work of [11], risk assessment methods are observed in four European countries and it is concluded that differences are mainly related to the factors included, methodology and risk types considered.

Thus, considering and the particularity of Brazil's food inspection system, to formally support the definition of suppliers to be visited by the inspection, this study aims to propose a multicriteria decision-making model, based on FITradeoff method, for selecting the suppliers to be visited. To this end, extensive research was conducted on the regulations to which milk and dairy production establishments are subject. Experts in the field were also consulted and a case study was conducted to demonstrate the usefulness and advantages of the proposed model.

2 Background

2.1 Managing Food Insecurity Risk

The importance of inspection focused on risk factors associated with foodborne diseases is highlighted in [3, 4]. For epidemiological surveillance to be effective, it is essential that health authorities clearly define these risk factors. This allows, through the investigation of outbreaks, the connections between diseases and their origins to be established, facilitating the identification and mitigation of risks associated with DTAs.

Risk-based inspection has become a constant worldwide, as for the [12], risk-based inspection aims to identify risk factors for protecting consumer health, determine priorities and allocate resources effectively and efficiently. Therefore, when hazards can have serious consequences for consumer health or when products have a greater risk of contamination, control must be intensified.

However, food risks have changed considerably in recent years and continue to change rapidly. The Canadian Food Inspection Agency (CFIA) has begun to evolve the way it manages risks and supports the industry's ability to compete globally by adopting technology to provide more efficient and responsive services. Therefore, it has committed to making better use of its data, reports and surveillance information to identify trends [13].

In view of the above, it is clear how risk-based inspection, based on the premise that all necessary measures for process control are implemented and that all safety risks associated with food products are being minimized during that process, has brought great advances in meeting safety and quality requirements [4].

Thus, institutional commitment, the dissemination of risk concepts among the entire professional team, and the adoption of safety and quality requirements throughout the production chain are essential for providing safer food to the consumer. In this context, the importance of Control Programs, like PNCRC of SFI stands out as modern instruments in risk management, undertaking its stages and using tools, with the objective of reviewing production processes and aligning them with food safety practices, preventing the occurrence of non-conformities and systematically reducing the prevalence of pathogens involved in food outbreaks [4]. In the next section, the problem studied will be defined, which is inserted in the PNCRC context.

2.2 Problem Statement

The PNCRC [5], implemented by MAPA, is a risk management strategy aimed at ensuring the safety of foods of animal origin produced in Brazil. Regarding milk and its derivatives, this plan includes monitoring the presence of a wide range of substances, such as veterinary drugs, pesticides, and environmental and industrial contaminants. When a non-compliance is detected (violation of the established limit), the program foresees the execution of corrective actions on rural properties, animal feed manufacturers, and slaughterhouses and processors.

Within the sampling plan, the PNCRC is scheduled and executed on an annual basis, although any emerging concerns may lead to its rescheduling throughout the year. The program consists of 4 (four) distinct but interrelated sampling lines: Monitoring Subprogram; Investigation Subprogram; Exploratory Subprogram; and Imported Product Control Subprogram [5].

The PNCRC Monitoring Subprogram includes random collections of samples from slaughtered animals and milk, eggs, honey, and fish sent for processing in establishments under SIF in response to collection orders (Official Analysis Requests - ROA) generated by the MAPA Central Unit. All establishments registered with DIPOA are subject to PNCRC sampling, including exporting establishments and those that sell their products only in the domestic market [5].

Sampling is managed through the computerized system called the Residue and Contaminant Control System (SISRES), through which ROAs are issued weekly. The system randomly defines, for each of the ROAs, the establishments in which the samples should be collected and indicates to the SIF user the matrix to be collected (animal tissue), the laboratory to which the sample should be sent, and the applicable deadline for collection. There is no defined sampling frequency per establishment. Collections are random, with variable frequency, without prior communication to the inspected party, which guarantees the surprise aspect of the inspection [5].

The definition of establishments where samples should be collected, although random, considers the size of the establishments. The greater the number of animals slaughtered and the volume of product processed in an establishment, the greater the probability of having a collection order assigned to it [5].

As recommended by Normative Instruction 77 of 2018, of the Ministry of Agriculture, Livestock and Supply, establishments producing milk and its derivatives must register their suppliers in an official system and keep them up to date. In addition, they must include in their self-control program: an updated registry of rural suppliers containing name, individual taxpayer registration number (CPF), address, daily volume, capacity, type and georeferenced location of the tank, lines, times and collection frequencies. Furthermore, according to Normative Instruction 77 of 2018, establishments under federal inspection, which have refrigerated raw milk, stored in individual or community refrigeration tanks, as well as milk received in cans must be collected for analysis in the laboratory of RBQL – Brazilian Milk Quality Network, with a minimum frequency of one sample per month, to assess the following parameters: fat content; total protein content; anhydrous lactose content; non-fat solids content; total solids content; somatic cell count; standard plate count and residues of veterinary products. RBQL must make the results of the analyses carried out available to the Ministry of Agriculture, Livestock and Food Supply, establishments and suppliers [14]. Milk samples submitted to PNCRC sampling are collected on rural properties, before mixing milk from different sources, to ensure that the sample refers to a single source [5].

As explained, MAPA holds various information on milk suppliers, as required by law. However, this information is not used to choose the milk supplier to be analyzed, within the context of the PNCRC. This choice is often not made based on the risk presented by the supplier. Thus, this work seeks to propose a model, based on the multicriteria approach, for identifying priority suppliers in the analyses carried out by the PNCRC. The aim is to structure the selection based on the risk associated with the supplier, the available resources and the time required for collecting and sending the sample to the defined laboratory. The multicriteria approach, presented in the next section, was used in this work to support this selection.

2.3 Multicriteria Approach

The use of a multicriteria approach in a decision problem is justified when the decision maker (DM), the individual to whom the power and responsibility for the decision is assigned, is faced with at least two alternatives for action and the choice of an alternative is guided by the desire to achieve multiple objectives, often conflicting with each other [15].

Although multicriteria decision support methods include the evaluation of alternatives, considering their performance in relation to multiple criteria, there are some divergences regarding the procedures adopted, depending on the purpose of the decision problem. Thus, [16] classifies decision problems into four types: choice or selection; ordering; classification and description.

Thus, several methods have been developed to support decision problems within the approach to multicriteria decision-making problems - MCDA and have been used in different contexts [17–22]. MCDA methods can be based on a single-criteria synthesis approach or an outranking approach. In the single-criteria synthesis approach, a single aggregation function is constructed, in which the Multi-Attribute Value Theory (MAVT) [23] stands out. In the outranking approach, the objective is to find an outranking relationship [24, 25] based on pairwise comparisons of the set of alternatives.

In this work, the problem of identifying suppliers to be inspected is characterized as a choice/selection problem, the objective of which is to define a set of suppliers that present a potential risk for production, considering milk samples already collected, volume of milk purchased from that supplier, and other issues. Analyzing the rationality of DM, the presence of tradeoffs when comparing different consequences was recognized, so that an MCDA method capable of dealing with tradeoff analysis is suitable for the model. In this context, FITradeoff for choice [26] was chosen to build the model, since it fits the characteristics of the problem, there is a system available with visual tools to assist in the elicitation of weights through a flexible and interactive process with linear programming problems that allow the use of partial information and require less time and number of questions to the DM [27] In the next section is presented the FITradeoff method, used in this work to identify suppliers considering these variables.

FITradeoff Method

The FITradeoff method [26, 28, 29] was chosen for the proposed model to support this problem. It supports multicriteria problems in the context of MAVT. Through a flexible and interactive process, this method allows obtaining and working with partial information about the DM's preferences. Thus, preference elicitation becomes less cognitively demanding for the DM, with less time and effort spent on the preference modeling process. Furthermore, the results are less inconsistent when compared to other multicriteria methods that require information on the criteria weights directly from the DM. These advantages reported in different applications [21, 30–32], added to the characteristics of the decision, justified the choice of this method.

Initially, the FITradeoff method was developed to support selection/choice problems in the context of a multicriteria decision [26]. Currently, variants of this method have been developed with the aim of supporting other types of problems: classification [33], ordering problems [34] and portfolio problems [35, 36].

The FITradeoff method for choice is based on a progressive reduction of the set of Potentially Optimal Alternatives (POA). In a multicriteria problem with n alternatives, an alternative a_i is considered potentially optimal if the global value of a_i, defined as the weighted sum, by the criteria weight, of the consequences of this alternative for each criterion, is greater than the global values of all other n-1 alternatives for at least one weight vector within the feasible weight space [26, 27]. In this sense, for an alternative a_i to be considered potentially optimal, the inequality in (1) must be valid for all $z = 1\ldots, n; z \neq i$ [28]:

$$\sum_{j=1}^{m} k_j v_j(x_{ij}) \geq \sum_{j=1}^{m} k_j v_j(x_{zj}) \tag{1}$$

In Eq. (1), k_j represents the scaling constant (or commonly called weight) of criterion $j\,(j = 1, \ldots, m)$ and $v_j(x_{ij})$ is the value of consequence x_{ij}, which consists of the evaluation of alternative a_i in criterion j, measured in a 0–1 scale according to the marginal value function of criterion j.

The FITradeoff mathematical model for choice problems seeks to verify the potential optimality of an alternative and is executed for all alternatives to form the subset of potentially optimal alternatives. At each interaction step, the following Linear Problem (LP) model is executed:

$$Max\, v(a_i) = \sum_{j=1}^{m} k_j v_j(x_{ij}) \tag{2}$$

$$s.t : k_1 > k_2 > \ldots > k_j > k_{j+1} > \ldots > k_m \tag{3}$$

$$\sum_{j=1}^{m} k_j v_j(x_{ij}) \geq \sum_{j=1}^{m} k_j v_j(x_{zj}), for\, all\, z = 1, \ldots, n; z \neq j \tag{4}$$

$$k_j v\left(x_j'\right) \geq k_{j+1} + \varepsilon \tag{5}$$

$$k_j v\left(x_j''\right) \leq k_{j+1} + \varepsilon \tag{6}$$

$$\sum_{j=1}^{m} k_j = 1 \tag{7}$$

Considering this LP model, if for at least one weight vector it is possible to maximize the global value of a_i (2), considering the current weight space according to the preference information given by the DM (3) and satisfying the potential optimality constraints in (4), then the alternative a_i is an POA for the problem. In the LP model, a small number ε is incorporated to make the inequalities in (5) and (6) computationally tractable in LP models.

At each interaction with the DM, the DM provides new preference information and the weight space is then updated with new inequalities of type (5) and (6), which are incorporated in the LP models, running again and searching for the updated subset of POA. The iteration steps continue until a unique alternative is considered optimal (i.e., this will be an optimal alternative for the problem); or until the DM feels satisfied with the actual set of potentially optimal alternatives [26, 28].

In the next section, the proposed model for selecting the set of suppliers to be inspected by SIF is presented.

3 Proposed Model

As a contribution of this research, the model presented in Fig. 1 is suggested for defining which set of suppliers should be inspected when the SIF visits milk and dairy producers.

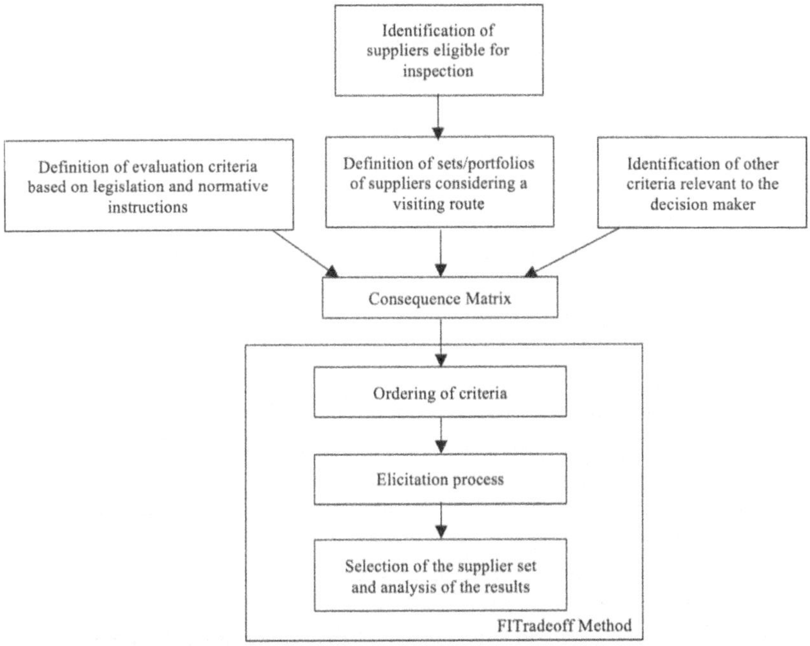

Fig. 1. Proposed model for supplier selection in the inspection process

The proposed model aims to present a practical and standardized way to select milk suppliers to be included in the analyses performed by the PNCRC, using the data available to the SIF. Thus, regarding the step "Definition of evaluation criteria based on legislation and normative instructions", the objective is to identify, based on legislation that already prioritizes the inspection of suppliers that pose a risk to this process, the criteria that allow suppliers to be evaluated in this direction. Depending on some circumstantial factors, the person responsible for inspection, who in this model has the role of DM, may include or even modify some criteria. For this reason, there is the stage "Identification of other criteria relevant to the decision maker". Most of the time, the DM comprises a single person (the inspector), but there are situations where a group of inspectors participate in this process.

As required by law, establishments (companies producing milk and dairy products) must provide RBQL laboratories with the information necessary to identify suppliers, including georeferenced location and quantification of the volume of milk produced. In addition, laboratories provide the results of analyses performed on samples from suppliers, sent by dairies [14]. Based on this information, sets of suppliers are defined to meet a viable visit route, since in addition to considering the potential risk, it is necessary to seek to optimize the inspection work and, therefore, the route that will be taken.

From this data, the consequence matrix is formed, which is necessary for the FITradeoff method to be started. The FITradeoff decision support system (DSS) is available on the website www.fitradeoff.org.br. It comprises three stages: ordering the criteria, eliciting the DM's preferences, and choosing the set of suppliers, along with the possibility of analyzing the results using the tools provided. It is important to emphasize that the role of the DM will be played by the SIF agent who is inspecting the establishment. An application of the proposed model is presented in the following case study.

4 Case Study

For this case study, data provided by a dairy company and entered into the SIF system were used. Considering the proposed model, presented in Fig. 1, and based on the location of 30 of the company's suppliers, 6 sets of suppliers were created that could be inspected in a single day of visits (an inspection route). These 6 sets constituted the alternatives to be evaluated in relation to the risk of presenting some type of contaminant, according to the PNCRC analysis scope.

Regarding the criteria used to assess potential risk, considering current legislation and the company's circumstantial issues, the criteria described in Table 1 were used.

Table 1 presents the criteria through which it is possible to identify the set of suppliers which present a greater risk and at the same time ensure better use of resources, which are scarce and limited.

In addition to these criteria, others can be considered to structure a multi-criteria decision-making approach capable of ensuring the identification of a set of suppliers which presents a greater risk of contamination. This increases the chances of more effective collection, provides opportunities for improving milk quality, and ensures standardization the process of selecting suppliers to be included in the scope of analysis of the PNCRC.

After defining the criteria and alternatives, it was possible to structure the consequence matrix using the model available on the FITradeoff website and following the instructions for filling it out, which are also available on the website. Considering the objective of the decision, the FITradeoff of choice was used, since the aim is to define, through a risk assessment, the best set of suppliers to be inspected.

Table 1. Description of decision criteria.

Criterion	Definition	Objective
Number of suppliers	Reflects the number of milk suppliers included in each set	Maximize
Somatic Cell Counting (SCC)	It assesses the quality of milk based on the somatic cell count per milliliter (SC/mL) in the set of suppliers	Maximize
Volume provided	The volume (in liters) of milk provided by the set of suppliers to the establishment	Maximize
Number of producers showing positive results for antibiotic by region	It reflects the presence of antibiotic residues, due to failures in monitoring the disposal of milk from cows undergoing antibiotic treatment, in milk samples	Maximize

After entering the data into the DSS, the stage of ordering the criteria began, according to their importance to the problem. In this stage, the DM, the person responsible for inspection, is asked about his/her preference regarding two fictitious alternatives A and B, which present different performances for any two criteria. An example of a question asked is shown in Fig. 2.

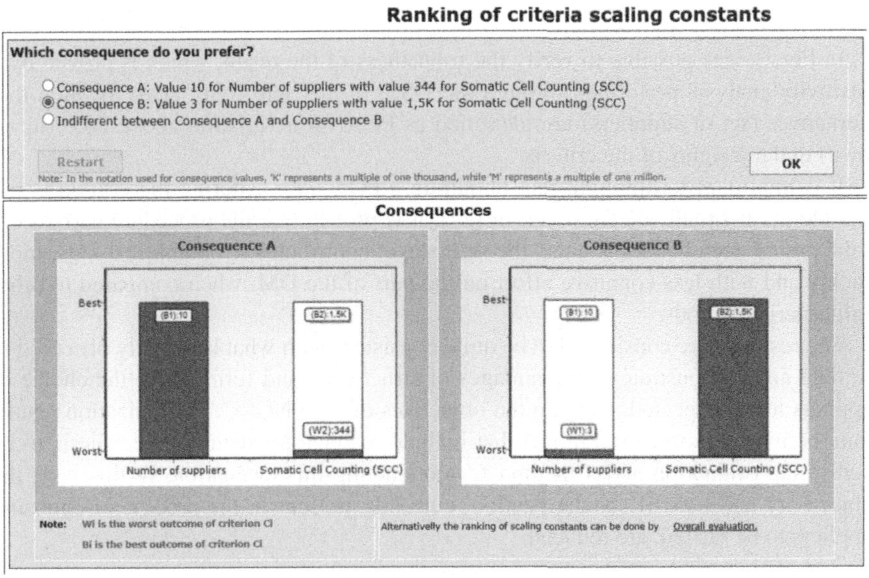

Fig. 2. Example of question in the ordering of criteria step

In this case study, after 25 interactions with the DM in this intra-criteria evaluation process, the following ordering of the criteria was defined based on their scale constants: Somatic Cell Counting (SCC) > Volume provided > Number of producers showing positive results for antibiotic by region > Number of suppliers.

This ordering of criteria was sufficient to identify a POA. This alternative is a group of 9 suppliers located nearby, which allows the tax authorities to visit all establishments. Together, these 9 suppliers are responsible for a daily supply of 2,850 L of milk. In milk samples collected from these suppliers by producers for registration with the SIF, the presence of somatic cells above the recommended level was identified, as well as the presence of antibiotics. The recommendation of this alternative of 9 suppliers, through FITradeoff, remains for the variation in the weights of the criteria presented in Fig. 3.

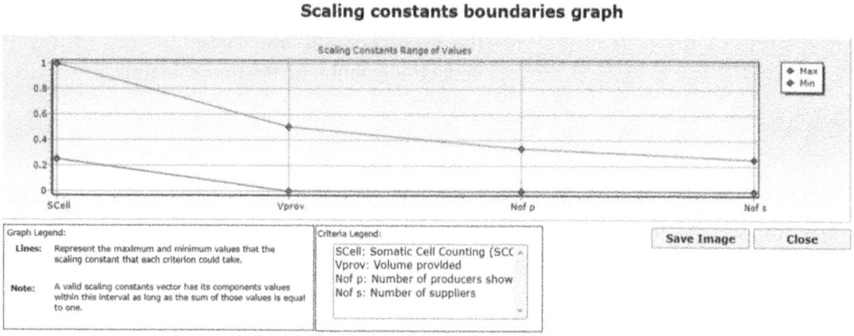

Fig. 3. Scaling constants boundaries for the alternative selected

In Fig. 3, it is possible to verify the robustness of the result, which is proven by a sensitivity analysis performed through the DSS. In this sensitivity analysis, the other alternatives (set of suppliers) are identified as POA for a variation above 30% (up or down) in the weights of the criteria.

It is important to highlight that obtaining a POA after ordering the criteria is not possible in all situations. However, in [27], it is proven that the procedure carried out in the preference elicitation stage, the next stage, contributes to obtaining a POA more quickly and with less cognitive effort on the part of the DM, when compared to other multicriteria methods.

The results were considered to be quite consistent with what is actually observed in the field and demonstrate the advantages of structuring and formalizing the choice of suppliers to be inspected, meeting the objectives of the PNCRC. Formalization results from the use of a structured model that includes a stage for defining the criteria to be used in the evaluation and a method for conducting this evaluation. In this way, the subjectivity and lack of standardization, currently present in the process of choosing suppliers to be visited, are reduced.

5 Conclusions

Food safety is an extremely relevant topic for society, since it involves public health. Representing 38% of establishments under federal inspection, milk and dairy production establishments are the most inspected sector, followed by the meat sector in Brazil. This data demonstrates the potential risk of this sector, which requires strict monitoring of milk and dairy production.

In this sense, the Brazilian Ministry of Agriculture, Livestock and Food Supply created the PNCRC to assess and minimize the risks of contaminants in food production. Although there is a formal methodology for defining visits to companies producing milk and dairy products, in the context of the PNCRC, there is no formal methodology for defining the suppliers to be included in the PNCRC when SIF inspectors visit the producing establishments. Therefore, it is up to the inspectors to define the suppliers to be visited, considering the data from the suppliers included in the SIF platform by the producing establishments, based on the milk samples collected.

Thus, in this study, a multicriteria decision-making model was proposed for selecting milk suppliers to be included in the milk analysis within the scope of the PNCRC, which is a risk management strategy focused on food safety. The proposed model sought to structure decision-making regarding the selection of a set of suppliers to be visited by the inspection. The FITradeoff SAD was used for this selection, considering the preferences of the DM (SIF inspector). When applying the proposed model, it was noted that the results obtained were quite concise and robust, demonstrating that the use of FITradeoff is extremely effective in this type of decision-making within the context of the PNCRC. Although a small set of alternatives was evaluated in the study, it is important to highlight that in contexts with more alternatives, studies [37, 38] demonstrate that the cognitive effort expended by the DM and even the number of interactions with the DSS, using FITradeoff, are smaller when compared to other methods that use the traditional tradeoff approach.

In future work, the possibility of using a multicriteria decision support model that already performs the stage of constructing supplier sets (supplier portfolios) can be evaluated, considering other important criteria and reducing the work of defining these sets by analysts and DM. Route optimization models can also be incorporated with the aim of more effectively defining these portfolios. Also, this work also meets the context of the PNCRC, which is a plan to prevent food contamination in Brazil. In other countries, similar inspection actions are being carried out, given the severity of the contamination in question. However, there may be differences regarding legal issues. Therefore, adjustments can be made to the proposed model, especially regarding the alternatives and criteria, to meet the specific issues of each country.

Acknowledgements. This study was financed in part by the Coordenação de Aperfeiçoamento de Pessoal de Nível Superior - Brasil (CAPES) – Finance Code 001. The authors thank the Foundation for Support of Science and Technology of Pernambuco (FACEPE) (grant number: PBPG-1312-3.08/22), which supported this study.

References

1. World Health Organization (2024) Food safety. https://www.who.int/news-room/fact-sheets/detail/food-safety. Accessed 10 Feb 2025
2. Liu, C., Mei, Y., Lei, Q., et al.: Fluorescent carbon dots based on food wastes: applications in food safety detection. Chem. Eng. J. **499** (2024)
3. Subedi, D., Paudel, M., Poudel, S., Koirala, N.: Food safety in developing countries: common foodborne and waterborne illnesses, regulations, organizational structure, and challenges of food safety in the context of Nepal. Food Front. (2024)
4. Olmedo, F.M.B., Omedo, L.M.B., Ribeiro, L.F.: Análise retrospectiva dos indicativos dos programas de monitoramento e controle microbiológico e físico-químico de alimentos do serviço de inspeção federal (2021)
5. Ministério da Agricultura. PNCRC: Manual Instrutivo do Plano Nacional de Controle de Resíduos e Contaminantes (2021)
6. Ministério da Agricultura. Anuário dos Programas de Controle de Alimentos de Origem Animal do DIPOA (2023)
7. Ministério da Agricultura. Cálculo do Risco POA: Manual para cálculo do risco estimado associado a estabelecimentos (2023)
8. Ford, J.H., Wilcox, C.: Quantifying risk assessments for monitoring control and surveillance of illegal fishing. ICES J. Marine Sci. **79** (2022). https://doi.org/10.1093/icesjms/fsac027
9. Röhrs, S., Rohn, S., Pfeifer, Y.: Risk classification of food incidents using a risk evaluation matrix for use in artificial intelligence-supported risk identification. Foods **13** (2024). https://doi.org/10.3390/foods13223675
10. Ze, Y., van Asselt, E.D., Focker, M., van der Fels-Klerx, H.J.: Risk factors affecting the food safety risk in food business operations for risk-based inspection: a systematic review. Compr Rev Food Sci Food Saf **23** (2024)
11. Borraz, O., Beaussier, A.L., Wesseling, M., et al.: Why regulators assess risk differently: regulatory style, business organization, and the varied practice of risk-based food safety inspections across the EU. Regul. Gov. **16** (2022). https://doi.org/10.1111/rego.12320
12. Food and Agriculture Organization of the United Nations. Risk-based food inspection manual. Food and Agriculture Organization of the United Nations (2008)
13. Canadian Food Inspection Agency. The Establishment-based Risk Assessment model for food establishments: the science behind it (2022)
14. Ministério da Agricultura P e A. Instrução Normativa N^{O} 77 (2018)
15. De Almeida, A.T.: Processo de decisão nas organizações: construindo modelos de decisão multicritério (2013)
16. French, S., Roy, B.: Multicriteria methodology for decision aiding. J. Oper. Res. Soc. **48** (1997). https://doi.org/10.2307/3010757
17. Alkraiji, A.I.: Weighting the challenges to the effectiveness of business intelligence systems in organisations: an empirical study of government organisations in Saudi Arabia. J. Decis. Syst. (2020). https://doi.org/10.1080/12460125.2020.1770436
18. Ferreira, D.V., de Gusmão, A.P.H.: Maturity assessment in the context of industry 4.0 - an application using FITradeoff method in a textile industry. In: Lecture Notes in Business Information Processing (2022)
19. Hajnić, M., Mileva Boshkoska, B.: A decision support model for the operational management of employee redeployment in large governmental organisations. J. Decis. Syst. **29** (2020). https://doi.org/10.1080/12460125.2020.1768681
20. Silva, M.M., Gusmão, A.P.H.D., Costa, A.P.C.S., Leyva-Lopez, J.C.: An outranking multicriteria method for nominal classification problems with minimum performance profiles. Math. Probl. Eng. **2019** (2019). https://doi.org/10.1155/2019/4078909

21. Ferreira, D.V., de Gusmão, A.P.H., de Almeida, J.A.: A multicriteria model for assessing maturity in industry 4.0 context. J. Ind. Inf. Integr. **38** (2024). https://doi.org/10.1016/j.jii.2024.100579
22. Silva, M.M., de Gusmao, A.P.H., de Andrade, C.T.A., Silva, W.: The integration of VFT and FI tradeoff multicriteria method for the selection of WCM projects. In: Conference Proceedings - IEEE International Conference on Systems, Man and Cybernetics (2019)
23. Keeney, R.L., Raiffa, H.: Decisions with multiple objectives: preferences and value trade-offs. IEEE Trans. Syst. Man Cybern. **9** (1979)
24. Brans, J.P., Mareschal, B.: The promethee VI PROCEDURE: how to differentiate hard from soft multicriteria problems. J. Decis. Syst. **4** (1995). https://doi.org/10.1080/12460125.1995.10511652
25. Pomerol, J.C., Roy, B., Rosenthal-Sabroux, C.: Developing an "intelligent" DSS for the multicriteria evaluation of railway timetables: problems and issues. J. Decis. Syst. **5** (1996). https://doi.org/10.1080/12460125.1996.10511691
26. De Almeida, A.T., De Almeida, J.A., Costa, A.P.C.S., De Almeida-Filho, A.T.: A new method for elicitation of criteria weights in additive models: flexible and interactive tradeoff. Eur. J. Oper. Res. **250** (2016). https://doi.org/10.1016/j.ejor.2015.08.058
27. de Almeida, J.A., Frej, E.A., Roselli, L.R.P., de Almeida, A.T.: Analytical aspects of combining holistic evaluation and decomposition elicitation for preference modeling in the FITradeoff method. Int. Trans. Oper. Res. (2024). https://doi.org/10.1111/itor.13470
28. de Almeida, A.T., Frej, E.A., Roselli, L.R.P., Costa, A.P.C.S.: A summary on FITradeoff method with methodological and practical developments and future perspectives. Pesquisa Oper. **43** (2023). https://doi.org/10.1590/0101-7438.2023.043spe1.00268356
29. de Almeida, A.T., Frej, E.A., Roselli, L.R.P.: Combining holistic and decomposition paradigms in preference modeling with the flexibility of FITradeoff. Cent. Eur. J. Oper. Res. **29** (2021). https://doi.org/10.1007/s10100-020-00728-z
30. de Morais Correia, L.M.A., da Silva, J.M.N., dos Santos Leite, W.K., et al.: A multicriteria decision model to rank workstations in a footwear industry based on a FITradeoff-ranking method for ergonomics interventions. Oper. Res. **22** (2022). https://doi.org/10.1007/s12351-021-00671-0
31. Ensinas, G., Pedreira, V., Luiz, M., et al.: Modelo multicritério baseado no vft e FITradeoff para análise do glp em grupos geradores (2023)
32. Martins, M.A., Garcez, T.V., De Gusmão, A.P.H., et al.: Multicriteria model based on FITradeoff method for prioritizing sections of Brazilian roads by criticality. Math. Probl. Eng. **2020** (2020). https://doi.org/10.1155/2020/8894402
33. Kang, T.H.A., Frej, E.A., de Almeida, A.T.: Flexible and interactive tradeoff elicitation for multicriteria sorting problems. Asia-Pac. J. Oper. Res. **37** (2020). https://doi.org/10.1142/S0217595920500207
34. Frej, E.A., de Almeida, A.T., Costa, A.P.C.S.: Using data visualization for ranking alternatives with partial information and interactive tradeoff elicitation. Oper. Res. **19** (2019). https://doi.org/10.1007/s12351-018-00444-2
35. Frej, E.A., Ekel, P., de Almeida, A.T.: A benefit-to-cost ratio based approach for portfolio selection under multiple criteria with incomplete preference information. Inf. Sci. (N.Y.) **545** (2021). https://doi.org/10.1016/j.ins.2020.08.119
36. Marques, A.C., Frej, E.A., de Almeida, A.T.: Multicriteria decision support for project portfolio selection with the FITradeoff method. Omega (U.K.) **111** (2022). https://doi.org/10.1016/j.omega.2022.102661

37. Frej, E.A., de Almeida, A.T.: Building a multicriteria decision model for solving a technical school location problem with the FITradeoff method. Pesquisa Oper. **44** (2024). https://doi.org/10.1590/0101-7438.2023.043.00281096
38. Vieira Dantas, N., Henriques de Gusmão, A.P., Teixeira Gonzaga, M.: A multicriteria model to support decisions regarding data protection compliance. J. Decis. Syst. (2023). https://doi.org/10.1080/12460125.2023.2263683

Blockchain Technology in Maritime Logistics: An Application in the Tactical Berth Allocation Problem

Ioannis Kypsidis, Magdalene Marinaki, and Yannis Marinakis(✉)

School of Production Engineering and Management, Decision Support Systems Laboratory, Technical University of Crete, Chania, Greece
{ikypsidis,mmarinaki,imarinakis}@tuc.gr

Abstract. Blockchain technology is evolving steadily and rapidly in this modern day, finding application in multiple sectors. While its primary application is in the financial sector, there is also a growing interest in supply chain management. Although research efforts recognize the need for further study into the correlation between these two sectors, the present research presents an important step towards understanding and exploiting this relationship. This paper analyzes the application of Blockchain in the supply chain, by first identifying the basic definitions and functions of the two concepts. Then, it examines their dynamics by analyzing the advantages and challenges accompanying Blockchain integration into supply chain processes. Through the presentation of practical examples and the evaluation of their impacts, the study highlights the importance of improving efficiency, reducing costs, and increasing transparency. The conclusion of the work highlights the prospects and challenges arising from integrating Blockchain into the supply chain (by solving the Tactical Berth Allocation problem in Maritime Logistics), while emphasizing the importance of further research and development in this area. As a result, the work provides a significant foundation for understanding how Blockchain technology can influence and transform the supply chain sector.

1 Introduction

The present paper delves into the interconnection between Blockchain technology and the supply chain. Initially, the work elucidates the definitions of Blockchain and the supply chain, followed by an explication of their general characteristics. Subsequently, the advantages and disadvantages of employing Blockchain are scrutinized. Finally, the paper examines the application of Blockchain in the supply chain, along with the research conclusions.

The paper underscores the evolution of Blockchain technology and its deployment in various sectors, particularly in the aftermath of the 2008 financial crisis, when Bitcoin emerged. Blockchain is a technological chain utilized for digital entries, which expands as new entries are added. The supply chain, on the other hand, is characterized as the process a product undertakes from its inception to its conclusion, i.e. from raw materials to the final consumer. The relationship between Blockchain and the supply chain is relatively recent but has garnered increased interest and adoption in recent years.

B. Delibašić et al. (Eds.): ICDSST 2025, LNBIP 546, pp. 43–55, 2025.
https://doi.org/10.1007/978-3-031-90863-7_4

To encapsulate the novelty and significance of the work, it is emphasized that, while Blockchain technology has predominantly developed in the financial sector, its application in the supply chain is a comparatively unexplored and nascent area. The paper endeavors to bridge this gap by offering a comprehensive analysis of the potential benefits and challenges of integrating Blockchain into the supply chain. Through this analysis, the paper significantly contributes to understanding how Blockchain technology may be used to improve supply chain processes by presenting viable answers to difficulties such as bottlenecks and other operational barriers.

This paper presents an innovative approach to the Tactical Berth Allocation Problem (TBAP) by integrating blockchain technology which improves transparency, security, and automation in the decision-making process. The use of blockchain technology allows for accurate and reliable data recording, as well as seamless information interchange among various stakeholders involved in port management. The application of blockchain in the TBAP and its impact on addressing issues such as berth allocation in ports are examined. The optimal allocation of ship berthing positions has a direct impact on port efficiency, hence TBAP arises as a critical problem in port management. TBAP arises due to the need for efficient management of port resources, particularly when demand for port services exceeds available capacity. In this context, the application of blockchain technology offers an innovative solution for automating the decision-making process by allowing precise and reliable data recording and transparent information exchange among involved parties. The use of smart contracts within blockchain enables automation and maximizes port performance while assuring process transparency and security. Blockchain in TBAP addresses the challenge of managing berthing positions for ships with greater efficiency, transparency, and security, improving port operational functionality and resource allocation.

This work is significant because it offers a new perspective on integrating blockchain into port management, approaching TBAP as a complex challenge that requires innovative solutions. By combining a technological approach with practical application in port management, this paper advances the understanding and development of blockchain as a tool for improving the supply chain and providing practical solutions to challenges in port management.

2 Literature Review

The term "Blockchain" consists of two components: the word "block" and the word "chain". It refers to a "chain" that is created and grows through "block" entries. Essentially, Blockchain is an ultra-modern environment through which monetary transactions can be reliably and securely conducted (Dutta, Choi, Somani & Butala, 2020). Blockchain can be considered as an online ledger, where users can easily and quickly execute transactions without the need for permission (Azzi, Chamoun & Sokhn, 2019). The term "Supply Chain" often refers to the management of the supply chain. More specifically, it is a group of procedures that assist in the production process of a product. That is, the path a product follows from raw materials to producers and, then, to the market and subsequently to the consumer (Mentzer, et al., 2001).

Blockchain is a chain that grows and expands according to the "blocks" that are transacted, i.e., the entries (Zheng, Xie, Dai, Chen & Wang, 2018). Blockchain first appeared as a concept in a book by Satoshi Nakamoto, which was about the use of the cryptocurrency Bitcoin (Nakamoto, 2008). Blockchain became synonymous with Bitcoin and began to have a significant influence, to the point where researchers' interest increased for further studies and, by extension, the usage of Blockchain in sectors such as health, supply chain, and so on expanded (Underwood, 2016).

The operation of Blockchain is analyzed through several stages. The first stage refers to the user's request for an entry. The second stage is the modification of the entry into a "block" form, followed by the search for a similar "block" and its attachment. Subsequently, there is the stage of validating the entries through an algorithm. Each new "block" is added to the file, with the aim of extending the chain (Bodkhe, et al., 2020).

The "block", i.e., the entries/transactions of Blockchain, consists of two parts, the body and the header. The body consists of two categories, (1) the transaction count and (2) the transactions. The number of transactions depends on the capacity of each "block" (Zheng, Xie, Dai, Chen & Wang, 2018). Moreover, the header consists of more categories: the block version (1), which reveals the rules that each "block" must follow, the block hash (2) and the Merkle hash (3). The first refers to the hashing of previous blocks, while the second to the hashing of transactions. The last categories relate to the timestamp (4), the nBits (5) and the Nonce (6) (Zheng, Xie, Dai, Chen & Wang, 2018).

It is important to articulate the types of Blockchain, as it is not a uniform body. More specifically, there are three types of Blockchain: a) permissionless, b) permissioned, and c) private chains (Dujak & Sajter, 2018). The first type, "permissionless," refers to block entries made without the approval of other users. The second type refers to a common decision of the users, where each member must give permission for it to be executed. Finally, the third is related to write permissions, which have a Blockchain recognition process (Yassein, Shatnawi, Rawashdeh & Mardin, 2019). Blockchain has some significant features that have propelled it to its present point. These features are presented as follows (Dutta, Choi, Somani & Butala, 2020, Niranjanamurthy, Nithya & Jagannatha, 2018, Zheng, Xie, Dai, Chen & Wang, 2018): Immutable, Transparent, Irreversible, Decentralized, Autonomous, Anonymous, Open-source, Provenance, Ownership and uniqueness.

In more detail, Blockchain is characterized as decentralized, as it allows users to monitor and be informed about the progress of transactions from multiple systems. Another characteristic is the autonomy and anonymity of Blockchain. The first refers to the ability of users to access blocks without interventions, while the second alludes to the blocks' anonymity (Niranjanamurthy, Nithya & Jagannatha, 2018). Furthermore, Blockchain is considered transparent and unique because the entries remain traceable in the online environment, and the blocks have their unique registration number (Dutta, Choi, Somani & Butala, 2020). Finally, there are the characteristics of irreversibility, i.e., the preservation of transactions in each file, and ownership, where each user's entry in the chain is provided with a document of ownership (Dutta, Choi, Somani & Butala, 2020).

There are three types of Blockchain classification systems: a) private, b) public, and c) consortium. The analysis of the aforementioned is based on certain categories (Monrat, Schelen & Andersson, 2019). Consensus is one of these categories. The consensus of the blocks is made by all users in the public Blockchain; in the consortium, some members are responsible for the consensus; and in the private, higher members choose who will be responsible for the consensus (Zheng, Xie, Dai, Chen, & Wang, 2018). There is also the read permission category, where on the public Blockchain, all users can read data, however, in the consortium and private, access is limited. Immutability is another category. Because of the larger number of members who can modify the data in the consortium and private Blockchains, it is almost impossible to modify transactions on the public Blockchain (Monrat, Schelen & Andersson, 2019). An additional category in the classification of Blockchain systems is efficiency. In the consortium and private Blockchain, there is better performance than in the public. Finally, there is the category of centralization, which is identified with the private Blockchain, as it selects its users and is governed by some higher members. The public is decentralized, as it has many independent members, and the consortium has some degree of centralization (Zheng, Xie, Dai, Chen & Wang, 2018).

It is now known that Blockchain has established itself in sectors other than cryptocurrency. Blockchain is applied in the economy, health, as well as the supply chain (Bodkhe, et al., 2016). **Smart Contracts**: Smart contracts are among the most modern applications of Blockchain, as they are electronic contracts that can be signed by two or more users. Once the contract is signed, the command is automatically executed. **Healthcare**: The adoption of Blockchain in the healthcare sector can aid with drug traceability. Traceability of medications is critical because there are several problems with their ingredients that cause health problems for users (Chang & Chen, 2020). **Economy**: One of the primary and essential sectors is the economy, which with the help of Blockchain began to recover, and electronic transactions made their appearance. Bitcoin, as a cryptocurrency, brought many changes to the value and transfer of money (Nakamoto, 2008). **Security**: Blockchain can also enhance the security sector, as it plays a key role in the security of electronic transactions (Chang & Chen, 2020). **Business**: Businesses have been upgraded rapidly with the help of Blockchain, due to financial services. Transactions are increasingly electronic, relieving businesses of asynchronous operations (Zheng, Xie, Dai, Chen & Wang, 2018).

Blockchain, as an idea, can bring many positive features and help organizations involved in the **supply chain** in several ways. However, this does not mean that there are no negative elements that need to be mentioned.

The absence of an intermediary is one of the positive features of Blockchain. Transactions are carried out without the use of a main administrator; instead, consensus programs are used to carry out the transactions simultaneously. Another positive aspect of Blockchain is the capabilities that each user has, both in control and in the information they have access to (Niranjanamurthy, Nithya & Jagannatha, 2018). Additionally, Blockchain provides users with integrity and reliability. Integrity contributes to the trust that Blockchain users can have in transaction processes, as they are carried out based on rules. Reliability contributes to the authenticity of longevity (Yassein, Shatnawi, Rawashdeh & Mardin, 2019).

Also, an important advantage of Blockchain is the stability and classification it provides. Entries are made in such a way that it is almost impossible to modify or remove them, and the data is distributed in a way that they can avoid any damages. Another key positive aspect of Blockchain is the lack of censorship and the ability to verify data.

That is, users can examine the accuracy of data transaction through verification. Also, it should not be overlooked that the Blockchain environment has free access to all users who want to participate (Erturk, Lopez & Yu, 2020). In addition, positive characteristics of Blockchain include the quality of the data, which is extremely high, and the fast transactions, which are carried out in just a few minutes, while in other cases they take days. An important advantage is also the low transaction cost in combination with the quality and speed (Niranjanamurthy, Nithya & Jagannatha, 2018).

As the advantages were mentioned earlier, there are also negative characteristics that Blockchain brings. A major disadvantage is the transaction process, as Blockchain databases have slower rates. This is mainly caused by the verification of signatures, consensus mechanisms, and transaction processing (Yassein, Shatnawi, Rawashdeh & Mardin, 2019). Additionally, another negative aspect of Blockchain is the security of browsing, which, while relatively high level, always has risks that cannot be addressed. It is estimated that a large percentage of Blockchain risks come from internet attacks to breach transactions (Erturk, Lopez & Yu, 2020). Also, significant disadvantages of Blockchain are the cost, i.e., the high energy use required for transactions, and the short period of time since its establishment. It takes time for Blockchain to start having more and more users who show trust (Niranjanamurthy, Nithya & Jagannatha, 2018). It should be noted that stability is considered a positive aspect of Blockchain, but this is not always the case. Changing data is quite difficult and often demanding and is frequently abandoned. Also, the use of "private keys" can cause difficulties in cases of absence. More specifically, if a private key is lost, the user's entries are automatically lost as well (Erturk, Lopez & Yu, 2020). Finally, the entries that can be made through Blockchain are of limited size and cannot be scaled in terms of information storage. This is also one of the most significant disadvantages of Blockchain (Niranjanamurthy, Nithya & Jagannatha, 2018).

3 Blockchain in Supply Chain

The application of Blockchain in the supply chain is quite modern and has started to develop in recent years with great success. Until today, this specific relationship is still being researched, since analysis studies are still being conducted (Tian, 2016). Two basic elements of Blockchain, which are significant factors for its application in the supply chain, are: a) secure transactions via Blockchain and b) automatic and verifiable transactions (Dujak & Sajter, 2018). According to the above factors, further areas of Blockchain application in the supply chain are developed.

Initially, there is the area that monitors the origin of the product and an area that generally monitors the product's journey. Also, there is the area of demand analysis and information provision for product transactions. Additional sectors are the automation of transactions and the reduction of environmental impacts (Dujak & Sajter, 2018). More specifically, the most known application of Blockchain in the supply chain is product traceability, i.e., the tracking of the product throughout its production process with the provision of valid location information (Dujak & Sajter, 2018). Furthermore, demand is referred to as a key area of Blockchain in the supply chain, as it refers to the data that exist for the demand and cost of each product. It contributes directly to the increase or decrease in profits (Dujak & Sajter, 2018). Also, the area of open user access to Blockchain chains in the supply chain can bring many positive features through the facilitation of bureaucratic matters (Dujak & Sajter, 2018).

There are several advantages of using Blockchain in the supply chain, which are presented as follows:

When compared to traditional manual work methods, Blockchain technology saves energy, time, and money. Also, with the application of Blockchain technology, security and compliance with rules in the working environment of the supply chain are enhanced. Moreover, ineffective communication affects unsatisfactory work. With the application of Blockchain, the above disadvantages begin to turn into advantages. Transactions are becoming more efficient, inexpensive, and safe as a result of the use of Blockchain in the supply chain (Dutta, Choi, Somani & Butala, 2020).

More advantages are transparency and time improvement. Transparency enhances product monitoring and distinguishes the hierarchical rights of the company. The usage of Blockchain in the supply chain improves delivery time when compared to traditional methods (Dutta, Choi, Somani & Butala, 2020).

Contrary to the advantages of using Blockchain in the supply chain, the disadvantages are minimal. The first is excessive use of technology. Speeds are reduced due to the excessive computers use. The second disadvantage is related to transaction data access. That is, there is a problem with determining which user who makes transactions should be assigned full access to the data.

Initially, a multitude of documents and transfers used in shipping can be facilitated using Blockchain, as there will be relief from both bureaucratic matters and unnecessary intermediaries. Regarding ports, which act as intermediaries in cargo transport, the use of Blockchain technology helps improve business activities.

Finally, in the transportation industry, the use of Blockchain for product tracking is a key part of transactions. Secure data transfer contributes to the elimination of problems in the supply chain (Yuan & Wong, 2016).

4 Tactical Berth Allocation Problem (TBAP)

The Tactical Berth Allocation Problem (TBAP) is a problem used in the field of shipping and port management (Bierwirth and Meisel, 2010). The objective is to decide the optimal berth allocation for unloading and loading ships while taking into account given parameters and conditions related to the ships and berths in a port. Let's analyze the process of the algorithm developed to solve this specific problem: First, we import the necessary libraries and read the data from an Excel file (data.xlsx). The data contains information about the number of berths, the number of ships, the capacity of the berths, the cost of using each berth, the capacity of the ships, the rate of arrival of the ships, and the demand of the ships. We create a Mixed Integer Programming (MIP) problem. We define the decision variables of the problem, which determine whether a ship is assigned to a berth. A value 1 indicates that the ship is assigned to the berth, while a value equal to 0 indicates that it is not assigned. We define the problem's objective function, which aims to maximize the usage of the berths by adding the products of the rate of arrival of the ships, the capacity of the ships, and the decision variables to the objective function. We set the constraints of the problem, which include:

1. Each ship must be assigned to only one berth.
2. Ships requiring refrigeration must be assigned to berths with cooling systems.
3. Each berth can be assigned to at most one ship.
4. The sum of the capacities of the ships assigned to a port/berth must not exceed the total capacity of it.

The number of berths used must not exceed the number of available berths.

We add a constraint for the total cost of the berths. This constraint ensures that the total cost of using the berths does not exceed the budget set in the "budget" variable.

4.1 Mathematical Modeling of the Problem

Effective space management in a port is a critical factor for improving its operation. In this work, we examine an optimization problem of assigning ships to berths to maximize the overall performance of the port. This optimization is carried out using a mixed-integer programming formulation. In this formulation, with N is denoted the number of ships, M is the number of berths, x_{ij} is a decision variable where it takes value equal to 1 if ship i is assigned to berth j and 0 otherwise, y_j is a decision variable that takes value 1 if berth j is used and 0 otherwise, *ship_capacity* is the capacity of each ship, *ship_rate* is the quantity of work that each ship needs and *ship_requirement* is a quantity of the specified requirements of each ship.

The objective function maximizes the overall performance of the port:

$$\max Z = \sum_{i=1}^{N} \sum_{j=1}^{M} ship\_rate_i * ship\_capacity_i * x_{ij}$$

The first constraint is that each ship is assigned to exactly one berth:

$$\sum_{j=1}^{M} x_{ij} = 1, \forall i$$

If a ship's cargo requires cooling, it is assigned to a berth with the capacity to cool the cargo:

$$x_{ij} \leq y_j, \forall i, \ with\ ship\_requirement_i = \prime refrigerated \prime$$

A berth can serve at most one ship:

$$\sum_{i=1}^{N} x_{ij} \leq 1, \forall j$$

The total capacity of the ships assigned to each berth must not exceed the berth's capacity:

$$\sum_{i=1}^{N} ship_{capacity_i} * x_{ij} \leq y_j * berth_{capacity_j}, \forall j$$

The total number of berths used must not exceed the total number of berths:

$$\sum_{j=1}^{M} y_j \leq M$$

The total cost must not exceed the budget:

$$\sum_{j=1}^{M} berth_{cost} * y_j \leq budget$$

4.2 Programs and Tools Used

In developing the algorithm for the application of blockchain in the supply chain, a wide range of tools and technologies were used, which were then combined to achieve our goals. Visual Studio, a powerful development environment, allowed the creation and programming of our algorithm in the Python programming language. This choice was made for the ease of syntax and the flexibility of the language for developing applications. In Python we use mainly the pandas library, a powerful tool for managing and analyzing data. It was used to process and handle data related to the supply chain, helping us to make data-driven decisions. Various tools and platforms were used for the integration of blockchain technology in the supply chain. Ganache is a blockchain simulation tool that creates a local environment for programming and testing our smart contracts. Truffle is a development framework that facilitates the creation, management, and programming of smart contracts in Solidity, the programming language of Ethereum. Solidity is the programming language used for developing smart contracts on Ethereum. It allows the expressive implementation of contract logic. Web3.js is a JavaScript library that enables interaction with the Ethereum blockchain from JavaScript applications, executing transactions, and retrieving data. With this integration of the above tools and technologies, we created an environment that allows effective problem-solving in the supply chain by leveraging blockchain technology.

Initially, for the solution of the optimization problem we use the PuLP library in Python. Initially, we read the data from Excel and, then, we transform the formulation of the Tactical Berth Allocation Problem in such a way that it is suitable for using the PuLP library and we use the Gurobi solver to find the optimal solution. Then, we started to interact with the Ethereum network and we will create Smart Contracts. To do this we will use the web3 library, which includes the packages you need to connect to the Ethereum network and create Smart Contracts. We, then, created a connection to a local deployment network accessible to a local address. This requires a virtual blockchain network, such as Ganache, to be active on the specified port and then, an Ethereum account is created using a private key.

The address variable is then assigned to a specific Ethereum address. An Ethereum address is a unique identifier for an account on the Ethereum blockchain. The balance variable is then assigned the value returned by the get_balance() function. This function retrieves the current balance of the specified Ethereum address. The balance represents the amount of Ether (the Ethereum network cryptocurrency) present in the account. The Ethereum address where the contract is deployed is assigned to the contract_address variable. An Ethereum contract is a self-executing program based on the blockchain. It can be thought of as a collection of data and code that can communicate with other accounts and other contracts on the network.

A JSON (Javascript Object Notation) file is used to load the contract_abi. The Application Binary Interface (ABI), defines the interface of the smart contract. It defines the functions, events and data structures that can be processed and interacted with from external data. ABI is essential for interacting with the contract, as it provides a formalized way of encoding and decoding function calls and data. In order to proceed we need to create a folder called "contracts" with the help of Truffle.

In summary, the first code performs the computation of the optimal distribution of ships on the berths of the port. The code uses the PuLP library to formulate and solve the mixed integer programming problem we have analyzed. The second code, on the other hand, uses the web3 library to interact with an Ethereum smart contract. This code allows connecting and interacting with the local blockchain and the smart contract developed on top of it.

5 Results

In the following, we can see the result of the development of the contract we created. To load the data into the blockchain, three blocks were used, and the cost of the process is indicated in the line with the total cost (see Fig. 1).

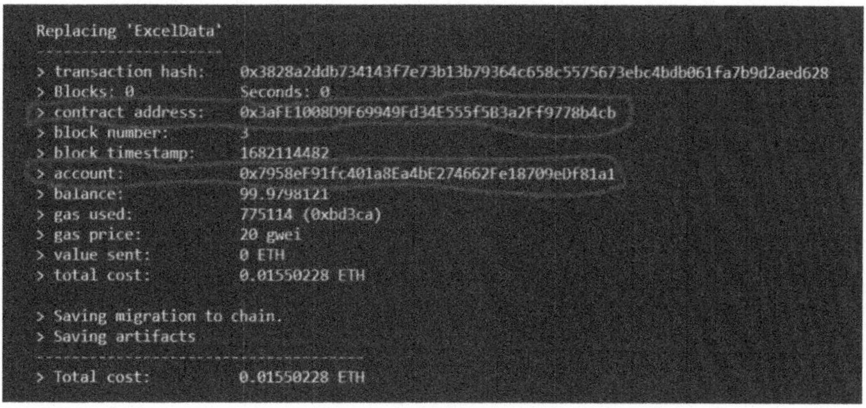

Fig. 1. Initial Blocks

Previously, we used the getRow function to allow users to receive data via the blockchain (see Fig. 2). This is useful for many reasons. For example, shipping companies can upload data for their vessels and cargoes, and port management can view the data without delay. Likewise, all parties involved can view the data with complete transparency.

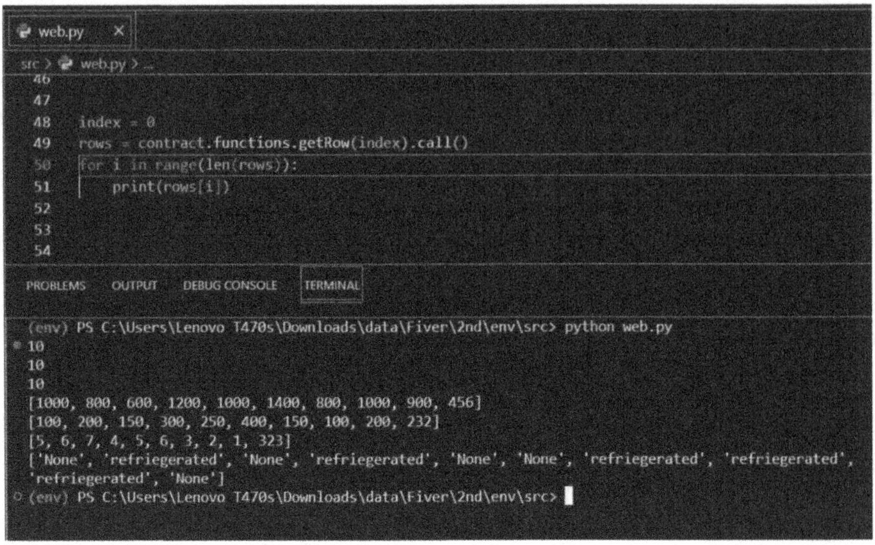

Fig. 2. Data from Blockchain

The Ganache user interface displays all available blocks in the blockchain (see Fig. 3). Each section contains transactions and information about the status of the network at any given time. The Ganache GUI displays these sections in a way that allows users to monitor network activity, review transactions and analyze the status of their blockchain.

Fig. 3. All available blocks in the blockchain

We transferred all transactions to the Ganache blockchain from Python (using the web3.py library) (see Fig. 4).

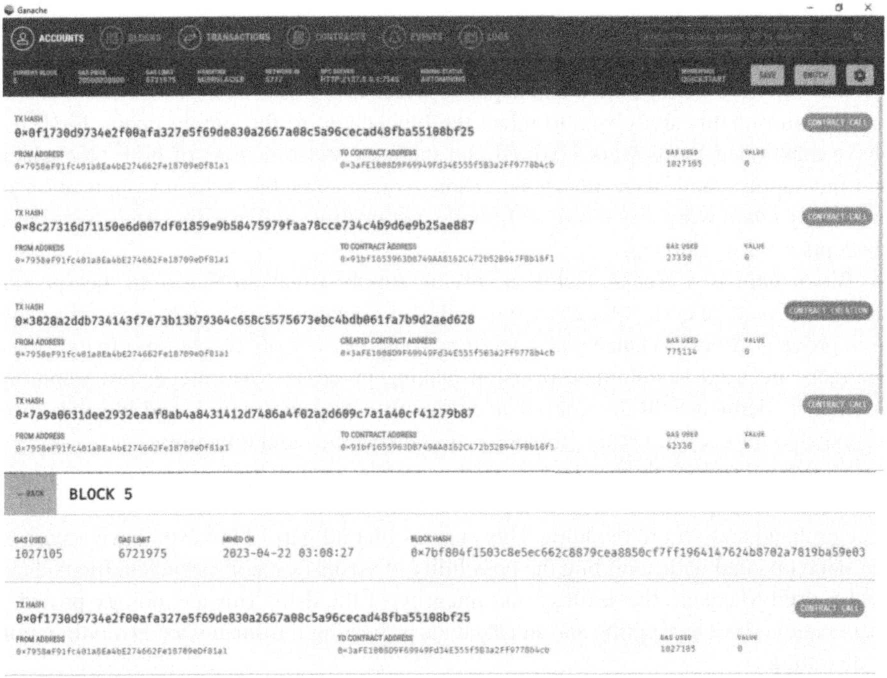

Fig. 4. Available Transactions

To summarize, we used the web3 library in Python to connect to the Ethereum network. We have connected to a local development network via the specified HTTP provider. By providing a private key, we created an Ethereum account and obtained the account address. We also queried the account balance, which indicates the amount of Ether in the account. We also interacted with a smart contract developed on the Ethereum network. We retrieved the ABI of the contract from a JSON file and created a version of the contract using the web3.eth.contract() function. This allowed us to read the contract's data and call its functions. We called the getRow() function with the index parameter using the appropriate contract object to retrieve a row of data from the contract's memory. The information was printed for further review after it was returned as a structured object. We also explored the creation of a Solidity smart contract. We created a contract called ExcelData that uses a structure called Row to store rows of data. Overall, we connected to the blockchain, interacted with a smart contract, retrieved data and performed various activities on the blockchain by using the web3 library and connecting to the Ethereum network.

6 Conclusions

Blockchain is a modern technological system that is becoming more and more prevalent in various sectors over time. Some of these sectors include healthcare, business and even the supply chain. The relationship between blockchain and the supply chain is being applied in organizations such as the shipping industry, public transportation, airports, etc. and can have both positive and negative effects.

The aim of this study was to relate the blockchain to the supply chain. First, the above-mentioned terms were clarified and the uses and functions of both blockchain and the supply chain were presented. Then, examples of blockchain's application in the supply chain were discussed, and finally, the positive and negative aspects of these concepts were mentioned.

Blockchain is a crucial factor in solving supply chain problems. In the specific problem of tactical berth allocation, blockchain is connected to the code in various ways to improve port performance and solve the problems of berth allocation. Firstly, smart contracts are used to automate decision-making processes. An integer programming problem is defined with the goal of maximizing port performance, and blockchain is responsible for executing this problem with transparency and reliability.

Specifically, blockchain solves the problem of data transparency and accuracy. Information about ship positions, port capacities and ship requirements are collected in a decentralized and secure medium. This ensures that all parties involved have access to the same updated data, avoiding the possibility of errors or inconsistencies. Blockchain is also used to ensure the security and integrity of the data. This technology provides secure methods of encrypting and storing data, protecting it from unwanted modification or alteration.

To summarize, blockchain is integrated into the problem of tactical berth allocation as a powerful tool to automate transparency and security in the decision-making process, improving the efficiency and overall operation of the supply chain.

References

Azzi, R., Chamoun, R.K., Sokhn, M.: The power of a blockchain- based supply chain. Comput. Ind. Eng. **135**, 582–592 (2019)

Bierwirth, C., Meisel, F.: A survey of berth allocation and quay crane scheduling problems in container terminals. Eur. J. Oper. Res. **202**(3), 615–627 (2010)

Bodkhe, U., et al.: Blockchain for industry 4.0: a comprehensive review. IEEE Access **8**, 79764–79800 (2020)

Chang, S.E., Chen, Y.: When blockchain meets supply chain: systematic literature review on current development and potential applications. IEEE Access **8**, 62478–62494 (2020)

Dujak, D, Sajter, D.: Blockchain applications in supply chain. SMART Supply Network, in EcoProduction, pp. 21–46, Springer (2018)

Dutta, P., Choi, T., Somani, S., Butala, R.: Blockchain technology in supply chain operations: applications, challenges and research opportunities. Transp. Res. Part E Logistics Transp. Rev. **142** (2020)

Erturk, E., Lopez, D., Yu, W.Y.: Benefits and risks of using blockchain in smart energy: a literature review. Contemp. Manag. Res. **15**(3), 205–225 (2020)

Mentzer, J., DeWitt, W., Keebler, J., Min, S., Nix, N., Smith, C., et al.: Defining supply chain management. J. Bus. Logist. **22**(2), 1–25 (2001)

Monrat, A.A., Schelen, O., Andersson, K.: A survey of blockchain from the perspectives of applications, challenges and opportunities. IEEE Access **7**, 117134–117151 (2019)

Nakamoto, S.: Bitcoin: A peer -to – peer electronic cash system (2008)

Niranjanamurthy, M., Nithya, B.N., Jagannatha, S.: Analysis of blockchain technology: pros, cons and SWOT. Clust. Comput. **22**, 14743–14757 (2018)

Tian, F.: An agri-food supply chain traceability system for China based on RFID & blockchain technology. In: 2016 13th International Conference on Service Systems and Service Management (ICSSSM). IEEE (2016). https://doi.org/10.1109/icsssm.2016.7538424

Underwood, S.: Blockchain beyond bitcoin. Commun. ACM **59**(11), 15–17 (2016)

Yassein, M.B., Shatnawi, F., Rawashdeh, S., Mardin, W.: Blockchain technology: characteristics, security and privacy, issues and solutions. In: 2019 IEEE/ACS 16th International Conference on Computer Systems and Applications (AICCSA). IEEE (2019)

Yuan, Y., Wong, F.Y.: Towards blockchain-based intelligent transportation systems. In: 2016 IEEE 19th International Conference on Intelligent Transportation Systems (ITSC). IEEE (2016)

Zheng, Z., Xie, S., Dai, H.-N., Chen, X., Wang, H.: Blockchain challenges and opportunities: a survey. Int. J. Web Grid Serv. **14**(4), 352–375 (2018)

Artificial Intelligence and Machine Learning

Examining Interface Preferences in a Decision Support System Using Eye-Tracking

Evanielle Barbosa Ferreira(✉), Lucia Reis Peixoto Roselli, and Adiel Teixeira de Almeida

Center for Decision Systems and Information Development (CDSID), Universidade Federal de Pernambuco, Recife, PE, Brazil
evanielle.ferreira@ufpe.br, {lrpr,almeida}@cdsid.org.br

Abstract. To support the decision-making process and enhance user interaction, Decision Support Systems (DSSs) are being developed and continuously improved. The aim of this study has been to investigate the interface of the FITradeoff DSS in order to investigate decision makers' preferences in holistic evaluation and decomposition process. This study analyzed the differences between four interfaces in a DSS, as well as eye movement data from 37 participants. In the experiment (total average duration of approximately 5 min), fixation and pupil data were captured using Area of Interest (AOI) drawings, as well as heatmaps and scan paths. The results revealed that the decision-makers seemed to spend more time, attention and cognitive effort on the screen declared as preferred. In addition, it is suggested that, in one of the interfaces, the change in the arrangement of the data on the graph resulted in an increase in the demand for the task, suggesting an increase in task time, duration of fixation and pupil diameter. Hence, based on the results, the study aims to improve the FITradeoff DSS providing modulations in the FITradeoff method and it DSS.

Keywords: Decision Support Systems (DSSs) · Behavioral experiment · Decision-making tasks

1 Introduction

Despite the studies on decision-making, there are still significant gaps in the use of neuroscience tools to investigate and modulate [1] Multi-Criteria Decision Making/Aiding (MCDM/A) methods and their Decision Support Systems (DSSs) [2], allowing for a greater understanding of the behavior of the decision-maker (DM) during the process.

There are several DSSs present in the literature to support multi-attribute decisions. Inserted in the context of Multi-Attribute Value Theory (MAVT), the Flexible and Interactive Tradeoff - FITradeoff method [3, 4] is implemented in a DSS, available at www.fitradeoff.org. The FITradeoff DSS enables DMs to express preferences in two ways: comparing consequences during the elicitation by decomposition and comparing alternatives in a holistic evaluation.

In the case of a DSS, much more than ensuring it has an aesthetically pleasing interface, it is essential to design an interface in which the proposed task is easy to

B. Delibašić et al. (Eds.): ICDSST 2025, LNBIP 546, pp. 59–75, 2025.
https://doi.org/10.1007/978-3-031-90863-7_5

understand, and which improves the DM's cognitive efficiency. [5] emphasize that user assessments regarding their perception of interfaces are produced automatically as a form of implicit assessments.

Therefore, this study investigates the design of the FITradeoff DSS. Based on the flexibility offered by the FITradeoff DSS, which allows Decision-Makers (DMs) to express preferences through two distinct paradigms, elicitation by decomposition and holistic evaluation [4], this study aims to investigate decision maker's preference regarding the DSS data presentation in order to improve the FITradeoff DSS. Hence, four different screens were created to illustrate the holistic evaluation using bar graphs, and the decomposition process, as discussed in Sect. 3.

Using psychophysiological data to investigate systems has paved the way for more efficient recognition of human patterns. Eye tracking, for example, has been used in various fields to measure people's visual behavior [6]. Using eye-tracking, several metrics can be extracted to assist in understanding the DM's behavior. This study analyzed view time, fixation time and pupil size for the screen area. Qualitative responses were also analyzed.

In this context, eye-tracking measures can be useful as objective techniques for studies on the layout of an interface. In this study eye movements were collected to support the investigation concerning a problem interface to solve the task of purchasing a cell phone.

Hence, the study contributes to the literature and to practice since (1) it provides analyses of eye movements in different types of interfaces of DSSs and (2) it provides evidence of the amount of attention and the cognitive process of interaction with interfaces. Hence, the aim of this study has been to investigate the interface of a DSS in order to suggest DMs preferences to improve the DSS constructed to implement the FITradeoff method [3, 4]. In addition, the study brings advances to the field of MCDM, filling the gap in the use of neuroscientific tools to investigate and modulate MCDM/A methods. [1]. Thus, behavioral aspects are collected through the use of eye tracking, allowing for a better understanding of the cognitive process and enabling modeling (transformations) in the DSS and providing powerful insights to improve decision-making processes [7].

The paper is structured as follows. Section 2 summarizes metrics and studies using eye movements. Section 3 describes the experiment, equipment and sample used in this study. Section 4 presents the results and discusses the findings of the analysis; and Sect. 5 presents the conclusions of this study and makes suggestions for future research.

2 Background

The period between 1970–1998 was marked by improvements in eye-tracking devices, allowing for more accurate and easily obtained measurements [8]. Eye-tracking can be used to analyze eye data and regions of visual attention at a given moment. Data obtained using an eye-tracker is free from the subjectivity of the participants, since the movements are captured without manipulation. In addition, the length and number of fixations are captured, making it possible to identify points on the screen that the user finds most interesting [9].

Saccades are rapid movements that the eyes make. Fixations, on the other hand, occur between saccades, when the eyes remain relatively still. By sequencing the movement of saccades and fixations, scan paths can be created.

Scanpaths with eyetracking have been analyzed for different purposes. The circular shapes represent the fixations, and the larger the circle, the longer the fixation in that region. The numbers in the circles indicate the sequence in which the fixation occurred and the lines between the circles represent the saccades [10].

Fixations can be attributed to different aspects such as cognitive tasks [11] and attention [12]. To [12], attention is a selectivity in perception, i.e. directing visual attention to a specific stimulus. Thus, in the decision-making process, if a particular stimulus does not obtain fixation, it is suggested that it is not available to the DM.

Regarding pupil diameter, several factors can affect this, such as changes in ambient light, color, spatial frequency, movement [13] and cognitive processes [14], as well as states of arousal, effort, interest and emotion [15–17].

There are many findings about eye movements in different tasks, and assumptions about cognitive processes. Starting with the paper by [18], it was clear that the presence of components influences the perception of the page. The authors aimed to identify which parts of web pages captured participants' attention the most by capturing their eye movements. Participants were asked to rate the visual attractiveness of 50 retail web pages, displayed in random order. They viewed six web pages (the three most liked and the three least liked) and rated the visual attractiveness. The results of the heat maps and analysis of fixations in the first five seconds helped to identify the components that probably influence the formation of users' opinions on the attractiveness of the pages.

In the study of [19], the authors conducted an experiment using eye movement capture to investigate the layout of agricultural product websites using three interfaces. The authors emphasize that the result of their study can serve as a reference for characteristic e-commerce sites and to reduce cognitive effort and improve the efficiency of products selected. [20] analyzed the task of selecting one of eight news items and showed the importance of position, saliency and topic in attracting and retaining attention and thus leading to selection.

Added to this, [21] investigated the effects of aspects of interface design on task performance of the elderly. For the authors, pupil diameter is proportional to the amount of information on the interface, meaning that a more transparent and understandable interface is conducive to reducing visual load. The study's findings revealed factors that influence task performance. In relation to the density of information in the interface, low density can simplify and reduce the cognitive load of the audience studied when they perform tasks.

Finally, [12] analyzed studies on eye movements specifically in decision-making and examined four theories. Some predictions were confirmed, but the authors point out that none were adequately explained. Among the insights, they suggested that attentional processes play an important part in decisions. Based on the findings in the literature, aspects derived from eye movements were examined during the decision-making process (Table 1).

Table 1. Metrics of this research

Analyzed aspect	Description	Study
Pupil diameter	Increased mental effort was associated with a dilated pupil The suggestion is that an increase in pupil diameter may result in more time being spent analyzing, understanding or pondering options	[22, 23]
Fixations	Fixation on a stimulus can enhance perception/attention Hence, attention probably causes two forms of after-effects in decision-making: 1) limiting the decision to the stimuli that have been fixated on and 2) increasing the influence of the information that has been fixated on	[12]
Time on Task	The suggestion for this aspect is that when DMs are faced with less preferred interfaces, they find it more difficult to compare information, and the task time increases significantly	

3 Description of the Experiment

To understand how DMs deal with various designs of screen in the process of comparing consequences, an eye-tracking experiment was carried out. The research investigated how people allocate visual attention when performing this task and their stated preference responses.

The survey was applied in the Federal University of Pernambuco (UFPE). The sample consisted of 37 participants. 19 male and 18 female, aged between 18 and 35. Moreover, 13 participants are matriculated on higher education courses, 6 participants have a bachelor's degree, 17 participants have a master's degree and 1 participant has completed a doctorate.

The experiment was conducted in the participants' native language. This approach was intended to avoid encouraging participants to evaluate the experiment in a different language, which could introduce additional cognitive effort unrelated to the problem itself but stemming from the language.

As a preliminary phase of the experiment, the participants were given information about the experiment and signed the Form of Informed Consent (FIC) required by the university's Research Ethics Committee (REC). Then, the participants were positioned so that the eye tracker could capture their eye movements, and the process of calibrating their pupils began. Finally, the fixation cross was displayed on a blank screen for 5000 ms.

During the experiment, participants were instructed on the multi-attribute decision problem of buying a cell phone. They had to choose between two consequences (A and B), evaluated on two criteria (Price - R$ and Memory - GB). In total, 4 screen options were created with the same problem, varying only the price range (R$ 3,500.00 to R$ 5,500.00) on some screens and the layout of these screens, i.e., the way this information was laid out, as shown in Fig. 1.

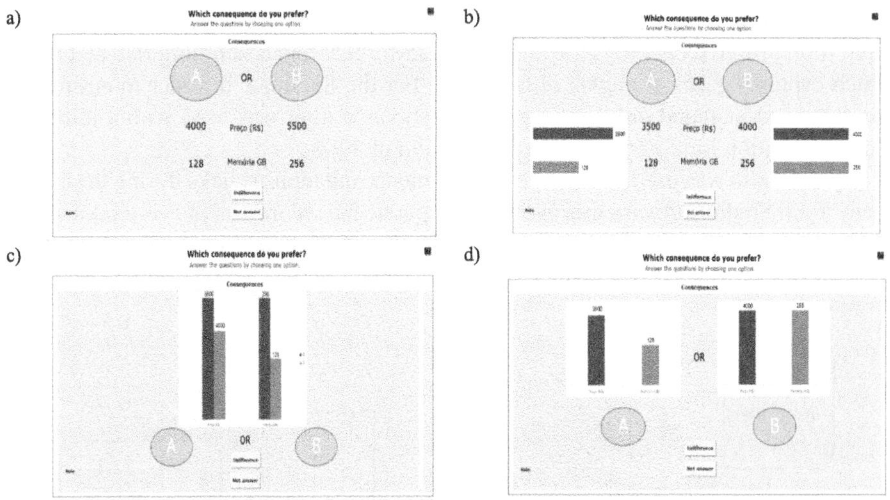

Fig. 1. Screens used in the experiment (a) Screen 1; (b) Screen 2; (c) Screen 3; (d) Screen 4

Screen 1 consisted only of consequence values arranged similarly to a table. In addition, each consequence had its own separate values (A and B). On screen 2, in addition to the information on screen 1, horizontal bar graphs were added. On screen 3, the information on the central values was removed and a vertical bar graph with joint consequence values was inserted, i.e., values for each criterion were grouped together in the same set of bars. Thus, the "Price" criterion was evaluated jointly, with consequence value A and consequence value B being compared. Similarly, the "Memory" criterion had consequence values A and B side by side.

Finally, on screen 4, similarly to screen 3, the information was presented graphically. However, the graph on screen 4 was made up of vertical bars and the performance of the alternatives was presented separately, i.e., the graph for consequence A showed A's performance on the "Price" criterion and on the "Memory" criterion.

The main objective in creating the four screens was to encompass the different data visualization options available in the FITradeoff DSS, which, during the presentation of the stages, offers various ways to display data. In the order of the criteria weights stage, the data is presented through pairwise comparisons in horizontal bar charts (like Screen 2). In the elicitation by decomposition stage, the data is presented in tabular form (like Screen 1) and in vertical bar charts (like Screen 4). Finally, in the holistic evaluation stage, vertical bar charts are used to display the performance of all alternatives across a single criterion (like Screen 3). Thus, the goal is to understand how these visualization methods are perceived and the preferences of the decision maker.

The screens were presented individually, without a predetermined duration. After evaluating a screen, the participant would proceed to the next one to choose the desired consequence. Screen 2 would then appear, and the process would be repeated until all 4 screens had been presented, as shown in Fig. 2. The experiment took approximately 5 min, and the same order was maintained for all participants.

Eye movements were recorded using a Tobii X-120 Eye Tracker from the manufacturer Tobii (Tobii Technology), with an accuracy of 0.5° and a sampling rate of 120 Hz, which captures gazes at angles of up to 35°. For the fixations, in order to ensure that the data used included only participants' fixations, a filter was used with a minimum fixation duration limit of 60 ms and a Threshold of 30°/s.

Tobii Studio was used to record eye movements and mouse clicks during the experiment. Tobii Studio software was used to incorporate the recordings of eye-tracking data. The software has a platform that enables the design of the experiment to be created, the recordings to be made and the data to be analyzed.

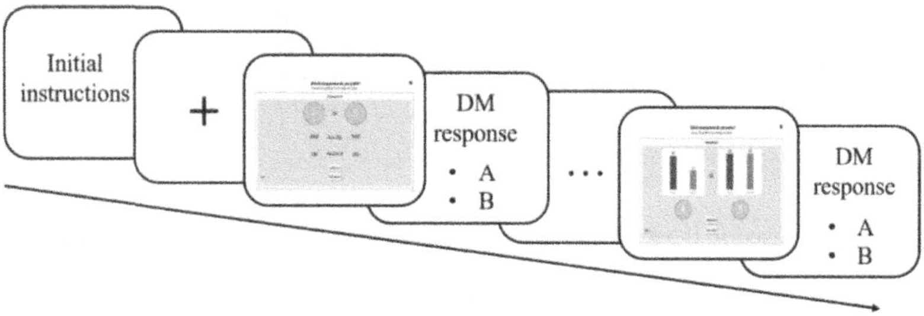

Fig. 2. Design of experiment

At the end of the experiment, the participant filled out a form with preference-related information, containing the following questions (Table 2):

Table 2. Preferential questions

1) Which of the screens did you like the MOST?	a) Screen 1 b) Screen 2 c) Screen 3 d) Screen 4
2) Which of the screens did you like the LEAST?	a) Screen 1 b) Screen 2 c) Screen 3 d) Screen 4
3) Which type of bar chart do you prefer (vertical bars or horizontal bars)?	a) vertical bars or b) horizontal bars
4) Which form of consequence visualization do you prefer (overall performance of alternatives by criteria or performance separated by alternative for each criterion)?	a) Overall performance or b) Performance separated

Using Area of Interest (AOI) drawings, the metrics used for the analysis in this study could be exported. The AOIs on the 4 screens included the areas with consequence values, the names of the criteria, the indifference and the did not respond buttons, the circles with consequences A and B, and for the cases where there were graphs, AOIs were added to the legend, data label and bars.

A few metrics were selected for analyzing eye movements: duration of fixation, average diameter of the pupil and duration of the analyzed interval. The results for these metrics obtained in the experiment are discussed in the next section.

4 Results and Discussion

After a manual inspection, only 37 out of 40 recorded signals were selected for further analysis. The rejected ones contained artifacts and a high noise level (i.e., data acquisition problems).

4.1 Eye-Tracking Analysis

To start the analysis, total time to do the task was evaluated for the four different types of screens presented since this can influence eye movement patterns (Table 3). Table 3 shows that the participants in the study spent between 12.90 and 18.13 s analyzing the screens until they entered the answer. The overall average time taken to complete the task was approximately 5 min (4.87 ± 1.36 min), considering the time taken for instructions, calibration and completing the final form. The screen 1 task took the longest to do, which was to be expected since it was the first screen presented, and it is natural for the DM to spend more time analyzing and understanding the first task.

It can be also observed that the duration of the task on screen 3 was closest to the duration of screen 1. When considering that it has the second highest mean pupil size (Table 3) and is the least preferred screen in the sample (Fig. 7), it is suggested that DMs, when dealing with less preferred interfaces, find it more difficult to compare information, resulting in a task taking longer to do. Longer reaction times are generally associated with increased task difficulty and greater pupil dilation (i.e., effort-based structure) [12]. In addition, this screen also had a high average number of saccades, a possible indication that the DM might have been searching for information on the screen (represented by the saccades). To complete this discussion, Fig. 3 shows the bee swarm visualization indicating where the DMs looked during a sliding window of time (5 to 7 ms). The focus of this visualization is to emphasize position rather than duration [24].

Table 3. Values from screens

	Interval duration (s)	Fixation duration (s)	Pupil size (mm)
Screen 1	18.13	14.26	3.456
Screen 2	12.90	9.67	3.372
Screen 3	17.56	13.48	3.379
Screen 4	14.57	10.63	3.346

In terms of fixation duration, the values for each of the AOIs analyzed can be seen in Tables 4, 5, 6 and 7. Note that DMs seem to use less cognitive effort on screen 2, followed by screen 4. An optimization of eye movement to reduce working memory is suggested, reflected by the reduction in the average number of fixations from screen 1 (2311) to screen 2 (1676) and from screen 3 (2046) to screen 4 (1796) that are required to make a decision.

Table 4. Fixation duration for Screen 1

AOI	128	256	4000	5500	A	B	Indifference	Not answered	Memory	Price	Total
FD (s)	0.202	0.171	0.931	1.017	0.741	0.511	0.418	0.147	0.659	1.168	14.256
%	1.4%	1.2%	6.5%	7.1%	5.2%	3.6%	2.9%	1.0%	4.6%	8.2%	

It is also suggested that the shorter fixation time may be related to becoming familiar with the interface, as screen 1 and screen 2 have the information in a central tabular form (despite screen 2 having the added graphic display), and screen 3 and screen 4 are both in the form of a bar graph. Thus, learning can also increase processing efficiency, decreasing the fixations.

The heat map expresses the degree of visual attention and comprises aggregated fixation data from the whole sample and is represented by colors, where red represents the highest level of attention used, while yellow and green represent a decrease in the level of attention [18]. Figure 4 shows the heat maps for each of the screens presented to the participants.

When analyzing the participants' aggregate heat map, it is clear that more attention is paid to the region of the consequence values for the "Price" criterion and its respective legend on screens 1 and 2. Thus, participants seem to spend more time analyzing, understanding or pondering this criterion in relation to the "Memory" criterion. In other words, the attention paid to the "Price" criterion may have had the effect of limiting the decision to the fixed stimuli (consequence values and the name of the criteria). In contrast, in the joint graph shown in screen 3, the DM seems to analyze the two criteria more homogeneously, a possible indication of similar weighting between both criteria and even in the legend.

Lastly, with regards to the interface shown in screen 4, consequence A seems to receive a higher level of attention related to the longer duration of fixation on the "Memory" criterion. Overall, the DM seems to pay attention to both the consequence values and the vertical bars.

Table 5. Fixation duration for Screen 2

AOI	128	128_graph	256	256_graph	3500	3500_graph	4000	4000_graph	A	B
FD (s)	0.143	0.041	0.202	0.007	0.580	0.103	0.725	0.106	0.365	0.241
%	1.5%	0.4%	2.1%	0.1%	6.0%	1.1%	7.5%	1.1%	3.8%	2.5%
AOI	Indifference	Not answer	Memory	Price	Bar_memory_right	Bar_memory_left	Bar_price_right	Bar_price_left	Total	
FD (s)	0.127	0.000	0.283	0.523	0.005	0.043	0.087	0.340	9.667	
%	1.3%	0.0%	2.9%	5.4%	0.1%	0.4%	0.9%	3.5%		

Table 6. Fixation duration for Screen 3

AOI	128	256	4000	5500	A	B	Indifference	Not answered
FD (s)	0.276	0.298	0.300	0.205	0.060	0.043	0.086	0.017
%	2.0%	2.2%	2.2%	1.5%	0.4%	0.3%	0.6%	0.1%
AOI	Legend	Memory	Price	Bar_memory_right	Bar_memory_left	Bar_price_right	Bar_price_left	Total
FD (s)	0.332	0.040	0.162	0.662	1.109	0.858	1.646	13.483
%	2.5%	0.3%	1.2%	4.9%	8.2%	6.4%	12.2%	

Analyses of attention have various repercussions in relation to decision-making. Fixations can reduce the demands on working memory [12]. In screen 2, for example, the graph inserted contains the same information as the central table on the screen but is simply a different visualization option for assessing the DM's preference of use during the task. The values for the duration of the fixation in the AOIs referring to the graphs on screen 2 were much lower than the AOIs in the table. It could be suggested that, by codifying that the information contained in the graph was the same as in the table, the DM decided to ignore the information relating to the new visualization and, in the end, to compare the consequences based on the previously familiarized option.

To finalize the analyses carried out, scan paths were generated for each of the participants as shown in the example in Fig. 5 and visual analysis of the scan paths was undertaken. It can be seen that the path taken is similar to the regions with the longest fixation times, as seen in the heat maps.

By analyzing visually each of the scan paths for the four screens, a series of conclusions can be drawn regarding the order of viewing and the elements that attract attention:

At a general level, the order of viewing was not identified. The gaze follows a different path between the participants, guided by the main AOIs, the consequence values of the "Price" criterion or buttons; in terms of direction, some of the sample tend to look following a vertical path initially (Fig. 5(a), (c)), followed by a more rectangular format when viewing the regions: a consequence value A for the "Price" criterion, a legend for the "Price" criterion a consequence value B for the "Price" criterion, a consequence value B for the "Memory" criterion, a legend for the "Memory" criterion, a consequence value A for the "Memory" criterion (Fig. 5(a)).

From the point of view of the general layout of the information on the interface, the central region of the screen seems to attract most of the attention, which is to be expected due to the clear definition of the areas with the information needed for decision-making in this region.

Table 7. Fixation duration for Screen 4

AOI	128	256	3500	4000	A	B	Indifference	Not answered	
FD (s)	0.160	0.167	0.177	0.171	0.056	0.056	0.087	0.003	
%	1.5%	1.6%	1.7%	1.6%	0.5%	0.5%	0.8%	0.0%	
AOI	Memory_right	Memory_left	Price_right	Price_left	Bar_memory_right	Bar_memory_left	Bar_price_right	Bar_price_left	Total
FD (s)	0.031	0.090	0.077	0.098	0.325	0.434	0.619	0.228	10.626
%	0.3%	0.9%	0.7%	0.9%	3.1%	4.1%	5.8%	2.1%	

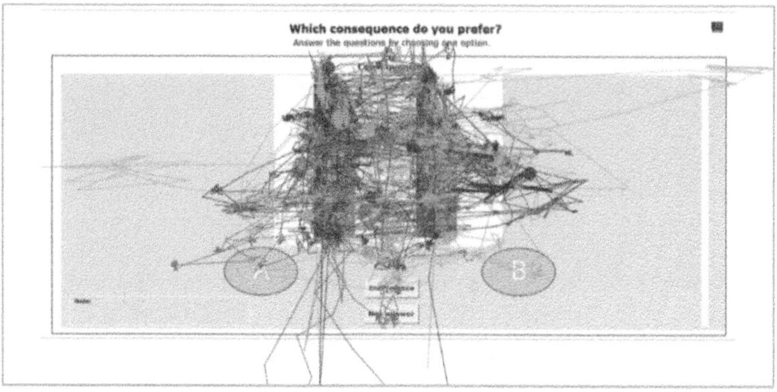

Fig. 3. Bee Swarm from Screen 3

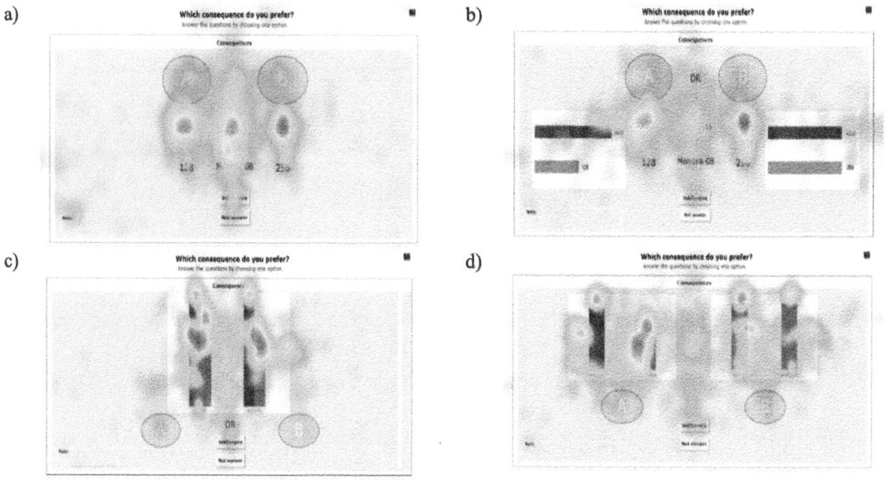

Fig. 4. Heat Map for (a) Screen 1; (b) Screen 2; (c) Screen 3; (d) Screen 4

On screen 2, consisting of both tabular (text-type information) and graphical information, the graphical visualization does not seem to have attracted the DM's attention during the task. On the other hand, to generate scan paths for the AOIs in the study, times were generated for the first fixation in each region of interest to understand what the scan path would be like considering only these regions. Figure 6 illustrates the number of times (using the number of participants) that each AOI was fixed for the first time. This made it possible to highlight the main AOIs for each position on each of the four screens presented. Furthermore, scan paths can be generated for each of the screens, aggregating data from each of the participants.

As decision theories discuss, goal-directed attention plays a significant role. In the present study, the DMs had the task of choosing a consequence in each of the screens presented. Thus, it is suggested that, as previously discussed, the DMs paid attention to important information or followed the effects of more useful information. Therefore,

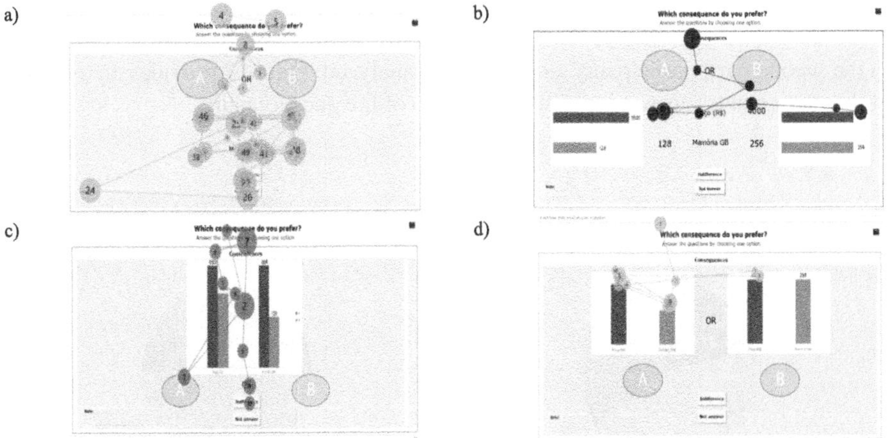

Fig. 5. Individual Scan Path of a participant for (a) Screen 1; (b) Screen 2; (c) Screen 3; (d) Screen 4

the scan path showed that most DMs start their analysis with the consequence values, followed by the criterion legend, and only a small proportion start on the right-hand side of the screen.

Screen 1				*Screen 2*				*Screen 3*				*Screen 4*			
Number of participants in each path															
37	36	34	30	37	36	33	33	37	37	35	34	37	35	31	22
Position				Position				Position				Position			
1st	2nd	3rd	4th	1st	2nd	3rd	4th	1st	2nd	3rd	4th	1st	2nd	3rd	4th
Number of participants for each AOI in this position															
A (16)	4000 (10)	Price (9)	5500 (11)	A (25)	3500 (9)	3500 (7), Price (6) and B (5)	3500 (7), 4000 (6) And Price (6)	Bar_price _left (18)	Bar_memory_left (8), Bar_price_right (7) and Bar_price_left (7)	Bar_memory_ left (9) and Bar_memory_ right (8)	Bar_memory _left (9)	Bar_price_left (10) and Bar_memory_ left (9)	Bar_price_right (9) and Bar_memory_l eft (7)	Bar_memory _right (10) and Bar_memory _left (6)	Bar_price _left (6)

Fig. 6. Positions to scan path only with AOIs

With regard to the scan paths created considering only the areas of interest in the study, a similar pattern could be suggested between screens 1 and 2, starting with the AOI containing the drawing with the name of consequence A, followed by the value of consequence A in the "Price" criterion, the legend of the "Price" criterion and the value of consequence B in the "Price" criterion.

As for screen 3, there is a predominance of paths starting with the left bar of the "Price" criterion, followed by the left bar of the "Memory" criterion. Finally, for screen 4, the suggested path is consistent with the type of graph and analysis required to complete the task. The suggested scanning path starts with the left bar of the "Price" criterion, followed by the right bar of the "Price" criterion, which makes sense since the consequence values are present in separate graphs on the screen.

4.2 Analysis of Declared Answers

In the second part, participants' answers were analyzed. Figure 7 provides an overview of the participants' answers to questions 1 to 4 of the questionnaire.

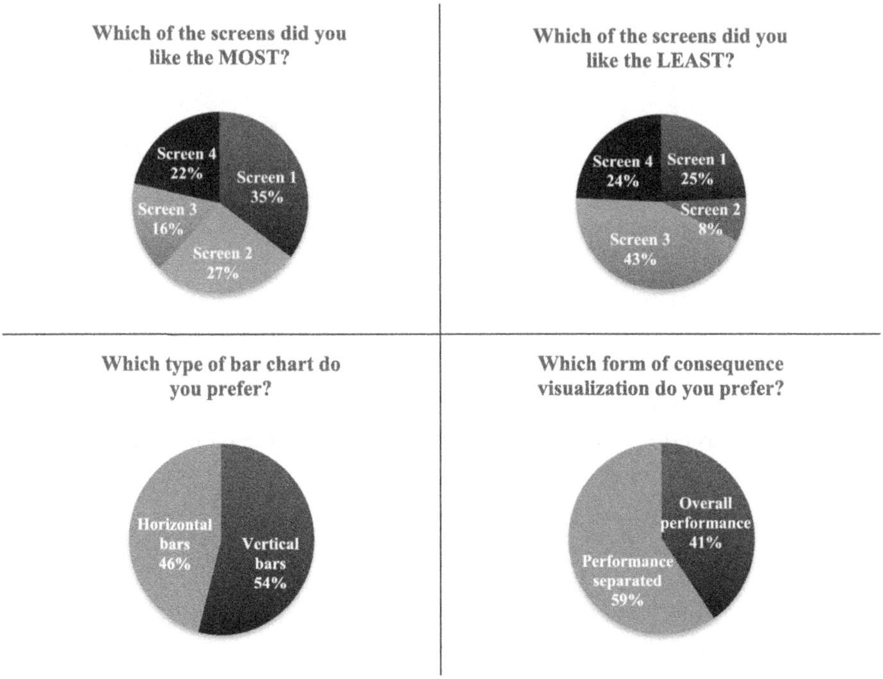

Fig. 7. Answers to questionnaire

In 13 out of 37 answers about screen preference (35%), participants indicated screen 1 as the most preferred screen. This screen consisted only of consequence values and buttons for choices. It is important to note that this screen had the longest average duration during the participants' analyses, as well as the longest fixation and pupil duration, as can be seen in Table 3. This suggests that participants used more cognitive effort and a higher level of attention on the preferred screen to analyze, understand or weigh up options. However, it is important to consider that, as it was the first screen presented to the participants, there is an initial exposure effect.

It is clear from Fig. 7 that screen 3 was the least preferred by the participants (43%). The identification of this non-preference for screen 3 can be supported by the shorter duration of fixation among all the screens presented, as shown in Table 3. Regarding the graphs, despite the non-preference for screen 3, the DMs declared a preference for the vertical bars (54%). However, they preferred to analyze the performance of the consequences separately (similar to screen 4).

Finally, Fig. 8 provides an overall summary of the key points obtained from the comparison of the four screens presented.

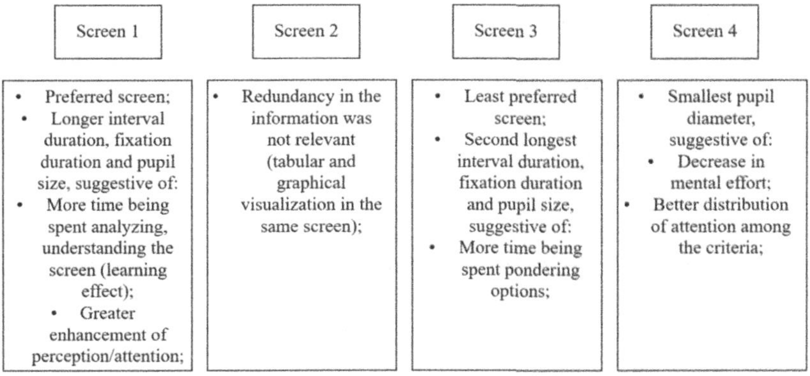

Fig. 8. Summary of the comparison of the four screens

5 Conclusion

In order to fill the gap observed in the literature concerning the use of neuroscience tools to investigate and improve MCDM/A methods [2], this study analyzed eye movement data and preference information on FITradeoff DSS interface. This study aims to capture non-controlled variables such as eye fixations to investigate DMs preferences in order to improve the design of the FITradeoff DSS, since most research deals with the subjectivity of DMs' preferences collected by surveys, for instance. On the other hand, aesthetic analyses alone do not guarantee that the decision task is easy to understand.

Therefore, this study used four different screens to analyze the interface during the decomposition process and holistic evaluation. To collect eye data, an experiment was carried out with thirty-seven people, using eye-tracking. All participants followed the same experimental procedure and had free time to view each of the screens presented. At the end of the experiment, they answered a form with preference questions, combining qualitative and quantitative analysis of the participants' preferences.

Using the heat maps, how the screens were fixed could be understood. In general, the acquisition of information was complete, despite some information being ignored by a few participants (not fixed). Based on results, it has been observed that DMs focused mainly on the "Price" criterion and its subtitle, with less focus on the other criterion of the problem (memory), especially in screens 1 and 2. In screens 3 and 4, there was a greater weighting between the criteria, illustrated by the duration of the fixation on the specific AOIs. With regard to the scan paths created considering only the areas of interest in the study, a similar pattern between screens 1 and 2 and a consistency in the suggested scan path on screen 4 could be suggested, according to the position of the elements on the interface.

Therefore, this study can suggest modulations for the FITradeoff DSS, for instance in order to consider pairwise comparison in the central part of the screen, such as Screen 1. The aim of this study has been to continue the modulations already done in the FITradeoff DSS [25].

Although the results and conclusions of this study are exciting, it is clear that these conclusions need to be validated with studies with a larger number of subjects, subdividing the sample with different orders of screen presentations and carrying out a statistical analysis of the data. In addition, there is the limitation of the diversity of criteria in the decision problem proposed in the experiment. For future studies, correlation between eye movements and decision marking preferences will be done.

Acknowledgment. This study had partial support from the Brazilian Research Council (CNPq) [grant 308531/2015-9; 312695/2020-9; 309237/2023-8] and the Foundation of Support in Science and Technology of the State of Pernambuco (FACEPE) [APQ-0484-3.08/17].

References

1. Wallenius, H; Wallenius, J.: Implications of world mega trends for MCDM research. In: Ben amor, S., de Almeida, A., de Miranda, J., Aktas, e. (eds.). Advanced Studies in Multi-Criteria Decision Making. Series in Operations Research, 1st edn., pp. 1–10. Chapman and Hall/CRC, New York (2020)
2. Hunt, L.T., Dolan, R.J., Behrens, T.E.: Hierarchical competitions subserving multi-attribute choice. Nat. Neurosci. **17**(11), 1613 (2014). https://doi.org/10.1038/nn.3836
3. Wang, X., shuang, Guo, F., & Li, M. ming.: Research on the neural mechanism of subconscious evaluation of mobile interfaces in smart apps. Int. J. Human-Comput. Interact. **40**(4), 1012–1021 (2022)
4. de Almeida, A.T., Almeida, J.A., Costa, A.P.C.S., Almeida-Filho, A.T.: A new method for elicitation of criteria weights in additive models: flexible and interactive tradeoff. Eur. J. Oper. Res. **250**(1), 179–191 (2016)
5. de Almeida, A.T., Frej, E.A., Roselli, L.R.P.: Combining holistic and decomposition paradigms in preference modeling with the flexibility of fitradeoff. Cent. Eur. J. Oper. Res. (2021). https://doi.org/10.1007/s10100-020-00728-z
6. Takeuchi, H., Habuchi, Y.: A quantitative method for analyzing scan path data obtained by eye tracker. In: CIDM 2007, pp. 283–286 (2007)
7. Linkov, I., Cormier, S., Gold, J., Satterstrom, F.K., Bridges, T.: Using our brains to develop better policy. Risk Anal. An Int. J. **32**(3), 374–380 (2012)
8. Duchowski, A.T.: A breadth-first survey of eye-tracking applications. Behav. Res. Methods Instrum. Comput. **34**, 455–470 (2002). https://doi.org/10.3758/BF03195475
9. Velásquez, J.D.: Combining eye-tracking technologies with web usage mining for identifying Website Keyobjects. Eng. Appl. Artif. Intell. **26**(5–6), 1469–1478 (2013)
10. Eraslan, S., Yesilada, Y., Harper, S.: Identifying patterns in eyetracking scanpaths in terms of visual elements of web pages. In: Web Engineering: 14th International Conference, ICWE 2014, Toulouse, France, Proceedings 14, pp. 163–180. Springer (2014)
11. Glöckner, A., Herbold, A.K.: An eye-tracking study on information processing in risky decisions: evidence for compensatory strategies based on automatic processes. J. Behav. Decis. Mak. **24**(1), 71–98 (2011)
12. Orquin, J.L., Loose, S.M.: Attention and choice: a review on eye movements in decision making. Acta Physiol (Oxf.) **144**(1), 190–206 (2013)
13. DiCriscio, A.S., Hu, Y., Troiani, V.: Task-induced pupil response and visual perception in adults. PLoS ONE **13**(12), e0209556 (2018). https://doi.org/10.1371/journal.pone.0209556
14. Daniels, L.B., Nichols, D.F., Seifert, M.S., Hock, H.S.: Changes in pupil diameter entrained by cortically initiated changes in attention. Vis. Neurosci. **29**(2), 131–142 (2012)

15. Hess, E.H., Polt, J.M.: Pupil size as related to interest value of visual stimuli. Science **132**, 349–350 (1960)
16. Kahneman, D., Beatty, J.: Pupil diameter and load on memory. Science **154**, 1583–1585 (1966)
17. Beatty, J.: Task evoked pupillary responses, processing load, and the structure of processing resources. Psychol. Bull. **91**, 276–292 (1982)
18. Djamasbi, S., Siegel, M., Tullis, T.: Generation Y, web design, and eye tracking. Int. J. Hum. Comput. Stud. **68**(5), 307–323 (2010)
19. Liu, L., He, X., Wan, F., Xiong, Z.: Study on usability of agricultural product web page layout based on eye tracker. In: 2016 International Conference on Automatic Control and Information Engineering, pp. 78–82. Atlantis Press (2016)
20. Navalpakkam, V., Kumar, R., Li, L., Sivakumar, D.: Attention and selection in online choice tasks. In: Masthoff, J., Mobasher, B., Desmarais, M.C., Nkambou, R. (eds) UMAP 2012. LNCS, vol. 7379, pp. 200–211. Springer, Heidelberg. (2012).https://doi.org/10.1007/978-3-642-31454-4_17
21. Zhou, C., Yuan, F., Huang, T., Zhang, Y., Kaner, J.: The impact of interface design element features on task performance in older adults: evidence from eye-tracking and EEG signals. Int. J. Environ. Res. Public Health **19**(15), 9251 (2022)
22. Kolnes, M., Uusberg, A., Nieuwenhuis, S.: Broadening of attention dilates the pupil. Atten. Percept. Psychophys. **86**(1), 146–158 (2024)
23. van der Wel, P., van Steenbergen, H.: Pupil dilation as an index of effort in cognitive control tasks: a review. Psychon. Bull. Rev. **25**(6), 2005–2015 (2018). https://doi.org/10.3758/s13423-018-1432-y
24. Tobii, A.B.: Tobii Pro Lab User Manual (Version 1.217). Tobii AB, Danderyd, Sweden (2023)
25. Barbosa Ferreira, E., Rani Soares de Vasconcelos, T., Reis Peixoto Roselli, L., Teixeira de Almeida, A.: Rebuilding a DSS with feedback from behavioral and neuroscience studies. Int. Trans. Oper. Res. (2025)

The Role of AI and Generative AI and Its Impacts on E-learning Processes and Outcomes: A System's View

Sean Eom(✉)

Southeast Missouri State University, Cape Girardeau, MO 63701, USA
sbeom@sbeom.edu

Abstract. While there is a growing body of research on using Artificial Intelligence technology and generative AI in education, the comprehensive examination of its impact on student's learning outcomes and satisfaction remains largely uncharted. We briefly introduce artificial intelligence and its subfields, including machine learning, deep learning, and generative AI. We then discuss the roles of AI in e-learning inputs, processes, and output through the lens of the System's view of the e-learning success model.

We examined, through the lens of the System's view of e-learning, the impact of AI on each of the isolated subentities, such as input (student motivation and instructor course facilitation), processes (self-regulated learning behavior, student-student dialogue, and student-instructor dialogue) and output (learning outcomes and satisfaction). We conclude that the systems approach helps us realize that the current literature on the role of AI and generative AI in e-learning systems failed to achieve the goal of e-learning systems as a dynamic set of interdependent subentities interacting to optimize learning outcomes and student satisfaction. We recommend conducting AI and generative AI research to contribute positively to learning outcomes and student satisfaction. To make that happen, we recommend going one step further by integrating AI/generative AI research results with structural equation modeling-based empirical research to optimize e-learning processes and learning outcomes.

Keywords: Artificial Intelligence · Generative Artificial Intelligence · System's View of E-learning Success Model · ChatGPT · Online Education

1 Introduction

E-learning research has evolved continuously with changing environmental variables, including information and communication technologies (ICT). ICT refers to diverse technologies that facilitate creating, moving, and communicating data/information among organizational members. Advances in emerging technologies, including ICT, have resulted in innovative changes in e-learning research. Among the various emerging technologies, Artificial intelligence (AI) has been considered one of the most promising technologies to facilitate and revolutionize students' e-learning outcomes and learner satisfaction [1–3].

B. Delibašić et al. (Eds.): ICDSST 2025, LNBIP 546, pp. 76–88, 2025.
https://doi.org/10.1007/978-3-031-90863-7_6

While there is a growing body of research on the use of AI technology in education, the comprehensive examination of its impact on student's learning outcomes and satisfaction remains largely uncharted. This research investigates the roles and effects of AI in facilitating students' learning processes, outcomes, and satisfaction in university online education through the System's view of the e-learning success model. There seems to be little research that provides a comprehensive review of opportunities and strategies for effectively utilizing ChatGPT in online university education.

There are two broad streams of discussions on the impact of AI on distance education. The first streams of research deal with AI topics, excluding generative artificial intelligence (GenAI). GenAI refers to a subset of AI that focuses on creating new content of text, images, sound, and other types of multimedia based on inputs received. Examples of GenAI include language models (GPT-3, GPT-4, and chatbots), image generation, music composition, video generation, 3D model generation, text-based storytelling, and code generation.

Chen et al.'s study [1] conducted qualitative research to investigate the effects of AI in administration, instruction (Improves learning experience, improves course contents, curricular developments – customized, individualized, personalized learning, and fostering instructional quality), and learning (tracking learning progress, developing intelligent learning systems, developing adaptive learning systems, enhancing academic integrity).

Numerous other similar research deals with the impact of all areas of AI techniques, including vision systems, natural language processing systems, expert systems, machine learning (ML) systems, and deep learning (DL) systems on similar or identical subjects (educational administration and learning [2, 4–9].

The second stream of research explicitly focuses on generative artificial intelligence, including students' perceptions of using ChatGPT in a physics class as a virtual tutor[10], using ChatGPT in a specific course [11], roles of Generative AI in promoting interactions among students [12], and encouraging active learning with ChatGPT [13].

The extensive review [13] shows that ChatGPT offers a variety of opportunities for higher education, including assessment innovation, instructional support, remote learning support, research design and development support, academic writing support, and administrative assistance and productivity. However, ChatGPT also presents several challenges and issues related to academic integrity, security, privacy, reliance on artificial intelligence, learning assessment, and information accuracy. The study offers a set of recommendations for the effective utilization of ChatGPT in higher education. It concludes that the application of ChatGPT in higher education presents benefits and challenges; thus, efforts and strategies are needed to ensure the effective use of ChatGPT for academic purposes.

Section 2 presents a brief introduction to artificial intelligence and its subfields. We then discuss the roles of AI in e-learning inputs, processes, and output through the lens of the System's view of the e-learning success model. The final section describes the conclusion and directions for future research.

2 A Brief Introduction to AI and Generative AI

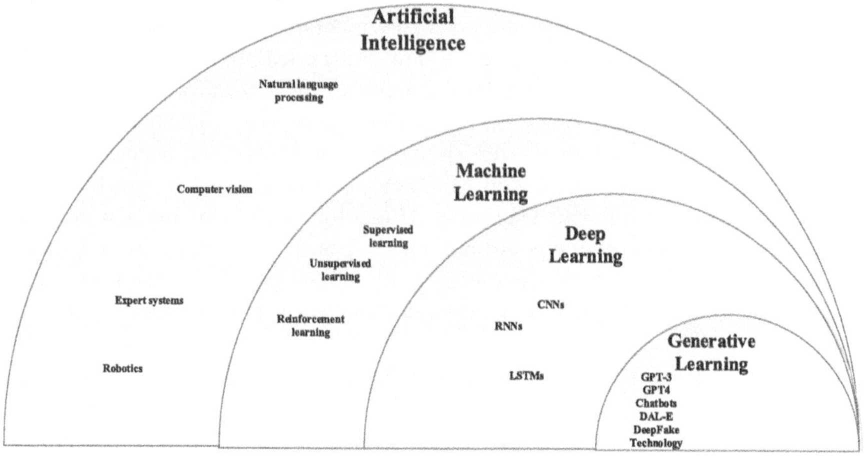

Fig. 1. Artificial Intelligence, Machine Learning, Deep Learning, and Generative AI

AI attempts to develop an intelligent computer system that simulates human behavior so that it can see (computer vision), speak (natural language procession, move (robotics), and think (expert systems). Its goal is the development of intelligent computer systems that perform tasks that require human intelligence. The AI field encompasses several subdomains: neural networks, expert systems, natural language processing, computer vision, robotics, intelligent agents, ML, DL, and Generative AI.

ML is a computer program aiming to "automate the task of analytical model building to perform cognitive tasks like object detection or natural language translation." [14, p.686]. ML algorithms allow computers to find hidden insights and complex patterns without being explicitly programmed. Instead, ML iteratively learns from problem-specific training data [15].

ML consists of several subfields: supervised learning, unsupervised learning, reinforcement learning, and deep learning. Supervised learning is a subfield of ML where models are trained on labeled data to make precise predictions or classifications [16]. Unsupervised learning unveils concealed pattern structures in unlabeled data [17].

DL is a subset of ML that uses neural networks to simulate the human brain's learning procedure. Moreover, DL is nested within the ML paradigm, a powerful subset characterized by neural networks sporting multiple layers. These layers facilitate the intricate processing and interpretation of data, unleashing AI's capability to comprehend complex patterns and glean profound insights [18]. "Deep" in deep learning refers to numerous hidden layers within the neural network. DL models are inspired by the structure and function of the human brain and consist of multiple layers of interconnected neurons. DL uses specialized architectures for different tasks: Convolutional Neural Networks (CNNs) for processing images and video; Recurrent Neural Networks (RNNs) for text or sequential data like time series; and Transformers for natural language processing [19].

2.1 Generative AI

Many higher education experts have described generative AI as one of the most disruptive technologies of our time. For example, Hsu and Ching [20, p. 603] provided the following statement:

> Generative artificial intelligence (GenAI), such as ChatGPT, has taken the world by storm. ChatGPT attracted 1 million users in 5 days and 100 million users in 2 months since its launch in November 2022.

It is a class of AI technologies that create content generate images with various types of multimedia (text, video, audio, and images) using ML and DL models such as supervised learning, unsupervised learning, and different types of neural networks as shown in Fig. 1 See [21] for a brief introduction to AI and its sub-elements.

Although generative AI has a broad capability of creating content in language models (GPT-3, GPT-4, and chatbots), image generation, music composition, video generation, 3D model generation, text-based storytelling, and code generation, OpenAI's Generative Pre-trained Transformer (GPT) has emerged as a groundbreaking tool that has begun to make significant strides in higher education, due to its potential to significantly restructure the educational landscape, altering the way students absorb, interact, and engage with academic content [22, 23].

ChatGPT and traditional chatbots have key differences in technology, context understanding, adaptability, and application areas. Readers are referred to [24] for detailed discussions on chatbots. ChatGPT is different from previous AI initiatives in several ways. (see Table 1) First and foremost, it can understand natural human language and comprehend, generate, and respond to complex queries and prompts. This capability of understanding human language is the basis of perceiving generative AI as a valuable tool with numerous benefits: personalized and immediate learning support, writing and brainstorming support, research and analysis support, visual and audio multimedia support, and administrative support [25].

3 Roles of AI on E-Learning Processes and Output Through the Lens of System's View of E-Learning Success Model

E-learning ecosystems are an open system of three entities: humans (students, instructors), learning management systems (LMS), and ICT that continuously interact with one another and with their interconnected environments and resources to optimize e-learning outcomes and student satisfaction (Fig. 2). The theoretical foundation of the model is based on constructivist learning theories. It is derived from a synthesis of the Virtual Learning Environment (VLE) effectiveness model of Piccoli et al. [27] and the framework of the Technology–Mediated Learning (TML) research of Alavi and Leidner [28].

The System's view of the e-learning success model provides a theoretically grounded conceptualization and incorporates more fully developed e-learning success measures. The model is built on three constructivist models (constructivism, collaborativism, and cognitive information processing model) [29, 30]. Furthermore, it is a systemic model with inputs, processes, and outputs over time and feedback loops.

Table 1. Comparison of ChatGPT and Chatbots

	ChatGPT (Generative and conversational AI-powered chatbot)	**Chatbots** (aka, Interactive agents, smart bots, and digital assistants)
Technology	*Cutting-edge technology (ML, DL, NLP) *Advanced AI language models developed by OpenAI using deep learning techniques	*Rule-based systems powered by ML, therefore, exhibit shallow conversational depth
Context Understanding	Hold more nuanced, context-aware conversations	Limited responses Less engaging interaction
Learning	Continuously learning from data	Requiring manual updates and programming
Adaptability	Highly adaptable and can improve over time	Less sophisticated adaptability
Range of Applications [25, 26]	*Broader tasks* *Interlocutor (conversational interactions) *Content provider (Creative Writing, Brainstorming ideas, ghostwriter) *Teaching assistant (tutoring) explaining concepts, providing examples *Evaluator (Grading student writing)	*Specific tasks* *Customer services *Booking systems *FAQs

3.1 Roles of AI for Supporting e-learning Inputs

The key elements of the e-learning ecosystem consist of two human entities (Instructors and students) and non-human entities (LMS and information communication technologies). Artificial intelligence is a subset of information and communication technologies.

Two human entities have many attributes, as listed in Fig. 2. Crucial attributes of students include motivation engagement efforts that critically affect the learning processes, resulting in student learning outcomes and satisfaction.

Recently, especially since the start of COVID-19, many universities have transitioned to online learning or hybrid modes. With this changing delivery mode, AI has become a catalyst that enables the e-learning process to proceed in a usually faster and more powerful way to improve learning outcomes and student satisfaction. The key elements of the e-learning ecosystem consist of two human entities (Instructors and students) and non-human entities (LMS and information communication technologies). Artificial intelligence is a subset of information and communication technologies.

Predicting Student Motivation Through ML and DL Algorithms
Student academic motivation is crucial to learning outcomes [31–33]. AI has significantly contributed to making it even more successful and effective in managing CSFs of distance learning systems. ML algorithms, including SVM and classification trees, and DL algorithms, such as artificial neural networks, analyze data such as grades, student-student and student-instructor interaction/dialogue, class participation, etc., to reveal student behavior and predict student academic motivation. Thus, it enables the e-learning system to provide timely interventions to a student who appears disengaged or frustrated. Further, the learning management system can offer motivational messages, suggest engaging content, or adapt the difficulty of assignments to maintain motivation levels [34, 35].

Enhancing Instructor's Roles using ChatGPT
The instructor is an indispensable entity in e-learning ecosystems. The fundamental question before discussing AI support for the instructor is defining the crucial roles of the instructor in distance learning. The systems view of the e-learning success model assumes that the instructor is only one element interconnected to other components in the e-learning ecosystem.

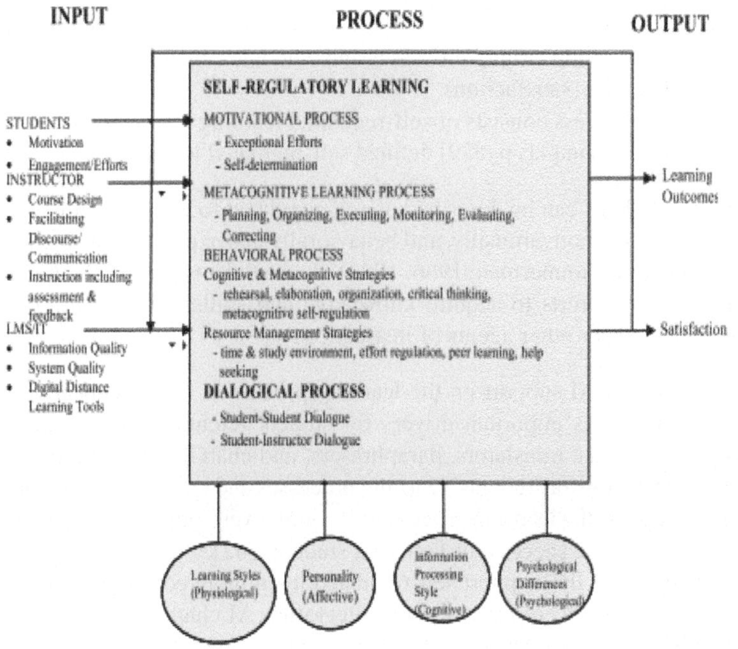

Fig. 2. A system's view of the e-learning success model Source: Eom and Ashill [29].

The study of [36] examined students' perceptions of instructors' roles in blended and online learning environments, using a total sample of 750 university students who participated in the study. Their confirmatory factor analysis identified five constructs representing essential instructor roles in e-learning and blended learning: course designer and organizer, discussion facilitator, social supporter, technology facilitator, and assessment designer. These findings are supported by many other research findings [37–40].

The roles of ChatGPT and chatbots support instructors in performing the following four complementary roles in university education [26].

*Interlocutor (conversational interactions): As an interlocutor, ChatGPT can perform roles as a conversational partner, providing relevant responses to users (the instructor and students.
*Content provider: ChatGPT produces written content and provides personalized feedback.
*Teaching assistant: As a tutor, ChatGPT provides personalized help and responds to student inquiries about course materials and assignments.
*Evaluator: as an evaluator, ChatGPT grades student writing, provides feedback, and analyzes student performance data.

3.2 Roles of AI for e-Learning Process Support

Learning outcomes are the outputs of the e-learning system, which takes two human inputs (the instructor and students) and various non-human inputs, such as LMS and CIT, to the transformation process. The technology-mediated learning (TML) framework consists of two significant inputs, instructional strategy, and information technology, that affect psychological learning processes and outcomes, including learners' affective reactions to e-learning (satisfaction).

The Learning Process consists of self-regulated learning behavior and dialogue and interaction. Zimmerman [41, p. 529] defined self-regulated learning as:

> In general, students can be described as self-regulated to the degree that they are metacognitively', motivationally, and behaviorally active participants in their own learning process (Zimmerman, 1986, 1989). Such students personally initiate and direct their own efforts to acquire knowledge and skills rather than relying on teachers, parents, or other agents of instruction.

The influence of AI support on the learning process in higher education has been increasing recently. As important drivers for student retention and learning success, generative AI tools like translators, paraphrasers, and chatGPT have become essential academic support tools for students' learning processes and knowledge creation. The use of AI tools to support the learning process is becoming very important and is considered a predictor of academic success for first-year students [42].

Chatbot and ChatGPT support self-regulated learning in education [43]. Three key pedagogical principles were proposed for integrating AI chatbots in classrooms using Zimmerman's self-regulated learning framework and judgment of learning (JOL) model [43]. Further, they advocate for incorporating goal setting (prompting), self-assessment and feedback, and personalization as three essential educational principles. They propose that teaching prompting is important for developing students' SRL.

Personalized intervention strategies utilizing SRL strategies were applied to those at-risk students to improve their learning performance [44]. The personalized intervention review activity not only helped students obtain higher learning performance but also prompted more significant improvements in the following learning strategies: rehearsal, critical thinking, metacognitive self-regulation, effort regulation, and peer learning.

The other important element of the process is interaction/dialogue. Eom and Ashill [29] defined dialogue as a positive and meaningful interaction. Their review of empirical studies on e-learning revealed that not every interaction leads to enhancing and/or increasing students' learning outcomes or satisfaction, as indicated by some studies. Thus, only meaningful interaction counts. Meaningful interaction directly influences learners' intellectual growth, stimulates learners' intellectual curiosity, and helps them engage in constructive learning activities that directly affect their learning outcomes.

Feedback, Student Engagement, and Student Learning Quality

Applying AI models to provide prompt and continuous feedback and improve student learning quality. This research integrated an AI performance prediction model with learning analytics visualization and feedback and conducted quasi-experimental research in an online engineering course to examine the differences in students' learning effects with and without the support of the integrated approach. The integrated approach increased student engagement, improved collaborative learning performances, and strengthened student satisfaction with learning [44].

Roles of AI in SS Dialogue

Many e-learning research studies have found that knowledge is socially and collaboratively constructed through sharing. Accordingly, interaction and dialogue between students and between the instructor and students are viewed as critical ingredients to the success of e-learning [29]. With the introduction of ChatGPT, "there is a global argument about whether AI tools can be seen as a new source of feedback for complex tasks. The answer to this question is unclear as there are limited studies and our understanding remains constrained" [45]. A recent study [45] used ChatGPT as a feedback source for students' argumentative essay writing tasks and compared the quality of ChatGPT-generated feedback with peer feedback. The results showed "ChatGPT provided more descriptive feedback including information about how the essay is written, peers provided feedback including information about identification of the problem in the essay" and that there are a potential complementary role for ChatGPT and students in the feedback process and the quality of the essays does not impact both ChatGPT and peer feedback quality.

3.3 Roles of AI for e-Learning Outputs (Learning Outcomes and Satisfaction)

Dropout Prediction and Minimization

The management of dropout in online education has been an important issue. Several factors contribute to e-learner dropout, including lack of motivation, lack of dialogue between students and the instructor, poor course design, etc. By analyzing historical LMS data and learning patterns, DL can predict potential obstacles and areas where students will likely struggle. With this insight, educators and platforms can provide targeted support and resources to help students overcome challenges before they lead to dropout. Personalized feedback, additional assistance, and adaptive content can make the learning experience more manageable and appealing, reducing the likelihood of abandonment [46].

Educational data mining is a valuable tool for predicting student performance. Machine learning predicts student performance by monitoring students' progress and identifying students at risk of failing the academic pathways. Quzdar et al. [47] proposed a machine-learning model using student data to make more precise predictions of students' performance and to identify struggling students. In addition, Alsharhan et al. [48] used three supervised machine learning methods (artificial neural networks, decision trees, and Naïve Bayes) to predict overall performance. They found that their ML model predicted student performance and those who needed intervention by testing several machine learning methods.

ML and DL are Enablers for the Development of Personalized Adaptive Learning Systems

AI and computer and information technologies have enabled an intelligent learning environment that facilitated the development of personalized learning (PL) and adaptive learning (AL). PL and AL are not identical but are often used interchangeably. King and South [49] defined PL as an approach in which the pace of learning and instructional content and its sequencing are optimized for the needs of each learner. On the other hand, Adaptive learning refers to technologies that dynamically adjust to the level or type of course content based on an individual's abilities or skill attainment in ways that accelerate learners' performance with automated and instructor interventions [50].

Personalization is a crucial aspect of modern e-learning platforms, and machine learning (ML) algorithms are vital in delivering personalized learning experiences to students. By leveraging data such as past performance, learning progress, and interaction patterns, ML algorithms can tailor learning content, pathways, and activities to suit individual needs, preferences, and learning styles. ML algorithms set the stage for personalized adaptive learning, a profound shift from one-size-fits-all education.

4 Conclusions

What we have seen in Sect. 3 was the impact of AI on isolated subentities such as input (student motivation, and instructor course facilitation), processes (SRL behavior, student-student dialogue, and student-instructor dialogue) and output (learning outcomes and satisfaction).

The systems approach to e-learning helps us view. It analyzes e-learning systems as a dynamic set of interdependent subentities interacting together, and e-learning systems are not explainable from characteristics of isolated subentities. Thus, a change in part of an interdependent system, such as student motivation, leads to ripple effects. In e-learning systems, a change in an input variable affects process variables, influencing learning outcomes and student satisfaction.

Our theoretical and practical contribution is that more research on AI and generative AI is needed to improve learning outcomes and student happiness. We suggest further maximizing the e-learning process and learning outcomes by combining empirical research findings based on structural equation modeling with generative AI research findings (Fig. 3).

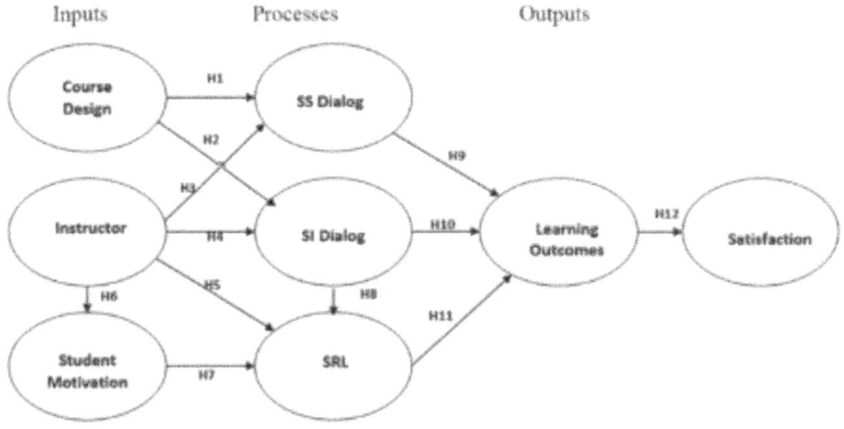

Fig.3. System's view-based research model (source: [32, p. 46])

An empirical study employing ChatGPT and LMS data to evaluate e-learning systems' efficacy is part of a recently developed research stream (e.g., [51]). In that study, structural equation modeling is used to empirically examine the relationship between learning outcomes and feedback (from both the teacher and ChatGPT). Students can get quick feedback on their schoolwork by using ChatGPT. Self-efficacy and emotional learning are significantly impacted by the combined use of feedback from traditional channels (teachers) and new generative AI (ChatGPT), and both mediators significantly influence learning results.

References

1. Chen, L., Chen, P., Lin, Z.: Artificial intelligence in education: a review. IEEE Access **8**, 75264–75278 (2020)
2. Ouyang, F., Zheng, L., Jiao, P.: Artificial intelligence in online higher education: a systematic review of empirical research from 2011 to 2020. Educ. Inf. Technol. **27**, 7893–7925 (2022)
3. Xia, Q., Weng, X., Ouyang, F., Lin, T.J., Chiu, T.K.F.: A scoping review on how generative artificial intelligence transforms assessment in higher education. Int. J. Educ. Technol. High. Educ. **21**(40) (2024)
4. Chen, X., Xie, H., Zou, D., Hwang, G.-J.: Application and theory gaps during the rise of artificial intelligence in education. Comput. Educ.: Artif. Intell. (2020)
5. Yang, C.C.Y., Chen, I.Y.L., Ogata, H.: Toward precision education: educational data mining and learning analytics for identifying students' learning patterns with ebook systems. Educ. Technol. Soc. **24**(1), 152–163 (2021)
6. Chen, X., Zou, D., Xie, H., Cheng, G., Liu, C.: Two decades of artificial intelligence in education: contributors, collaborations, research topics, challenges, and future directions. Educ. Technol. Soc. **25**(1), 28–47 (2022)
7. Wongvorachan, T., Lai, K.W., Bulut, O., Tsai, Y.-S., Chen, G.: Artificial intelligence: transforming the future of feedback in education. J. Appl. Test. Technol. **23**(1) (2022)
8. Chen, X., Zou, D., Cheng, G., Xie, H., Jong, M.: Blockchain in smart education: contributors, collaborations, applications and research topics. Educ. Inf. Technol. **28**(4), 4597–4627 (2023)

9. Wang, S., Sun, Z., Chen, Y.: Effects of higher education institutes' artificial intelligence capability on students' self-efficacy, creativity and learning performance. Educ. Inf. Technol. **28**(5), 4919–4939 (2023)
10. Ding, L., Li, T., Jiang, S., Gapud, A.: Students' perceptions of using ChatGPT in a physics class as a virtual tutor. Int. J. Educ. Technol. High. Educ. **20**(1) (2023)
11. Ellis, A.R., Slade, E.: A new era of learning: considerations for chatGPT as a tool to enhance statistics and data science education. J. Stat. Data Sci. Educ. **31**(2), 128–133 (2023)
12. Liang, J., Wang, L., Luo, J., Yan, Y., Fan, C.: The relationship between student interaction with generative artificial intelligence and learning achievement: serial mediating roles of self-efficacy and cognitive engagement. Front. Psychol. **14**(1285392) (2023)
13. Jayasinghe, S.: Promoting active learning with ChatGPT: a constructivist approach in Sri Lankan higher education. J. Appl. Learn. Teach. **7**(2) (2024)
14. Janiesch, C., Zschech, P., Heinrich, K.: Machine learning and deep learning. Electron. Mark. **31**, 685–695 (2021)
15. Bishop, C.M.: Pattern recognition and machine learning. Springer, New York (2006)
16. Gupta, V., Mishra, V.K., Singhal, P., Kumar, A.: An overview of supervised machine learning algorithm. In: Proceedings of the 11th International Conference on System Modeling & Advancement in Research Trends (SMART), Moradabad, India (2022)
17. Talaei Khoei, T., Kaabouch, N.: Machine learning: models, challenges, and research directions. Future Internet **15**(332) (2023)
18. LeCun, Y., Bengio, Y., Hinton, G.: Deep learning. Nature **521**(7553), 436–444 (2015)
19. Alom, Z., et al.: A state-of-the-art survey on deep learning theory and architectures. Electronics **8**(292) (2019)
20. Hsu, Y.-C., Ching, Y.-H.: Generative artificial intelligence in education, part one: the dynamic frontier. TechTrends **67**(4), 603–607 (2023)
21. Eom, S., Amroune, M., Khediri, A., Laour, M.R.: Roles of machine learning and deep learning for supporting E-learning processes and learning outcomes through the lens of system's view of e-learning success model. In: Book Roles of Machine Learning and Deep Learning for Supporting E-Learning Processes and Learning Outcomes through the Lens of System's View of E-Learning Success Model (2024)
22. Jo, H.: From concerns to benefits: a comprehensive study of ChatGPT usage in education. Int. J. Educ. Technol. High. Educ. **21**(1) (2024)
23. Fauzi, Tuhuteru, L., Sampe, F., Ausat, A.M.A., Hatta, H.R.: Analysing the role of ChatGPT in improving student productivity in higher education. J. Educ. **5**(4), 14886–14891 (2023)
24. Adamopoulou, E., Moussiades, L.: Chatbots: History, technology, and applications. Mach. Learn. Appl. **2**, 100006 (2020)
25. Chan, C.K.Y., Hu, W.: Students' voices on generative AI: perceptions, benefits, and challenges in higher education. Int. J. Educ. Technol. High. Educ. **20**(1) (2023)
26. Jeon, J., Lee, S.: Large language models in education: a focus on the complementary relationship between human teachers and ChatGPT. Educ. Inf. Technol. **28**(12), 15873–15892 (2023)
27. Piccoli, G., Ahmad, R., Ives, B.: Web-based virtual learning environments: a research framework and a preliminary assessment of effectiveness in basic IT skills training. MIS Q. **25**(4), 401–426 (2001)
28. Alavi, M., Leidner, D.E.: Research commentary: technology-mediated learning – a call for greater depth and breadth of research. Inf. Syst. Res. **12**(1), 1–10 (2001)
29. Eom, S.B., Ashill, N.: The Determinants of Students' Perceived Learning Outcomes and Satisfaction in University Online Education: An Update. Decis. Sci. J. Innov. Educ. **14**(2), 185–215 (2016)
30. Eom, S.B., Ashill, N.J., Arbaugh, J.B.: Guest editors' introduction to the special issue. Decis. Sci. J. Innov. Educ. **14**(2), 124–127 (2016)

31. Eom, S.B., Ashill, N., Wen, H.J.: The determinants of students' perceived learning outcome and satisfaction in university online education: an empirical investigation. Decis. Sci. J. Innov. Educ. **4**(2), 215–236 (2006)
32. Eom, S.B., Ashill, N.: A system's view of e-learning success model. Decis. Sci. J. Innov. Educ. **16**(1), 42–76 (2018)
33. Ateş, H., Köroğlu, M.: Online collaborative tools for science education: boosting learning outcomes, motivation, and engagement. J. Comput. Assist. Learn. (2024)
34. Babić, I.Đ: Machine learning methods in predicting the student academic motivation. Croat. Oper. Res. Rev. **8**(2), 443–461 (2017)
35. Wang, X., Zhang, Y., Yu, S., Liu, X., Wang, F.-Y.: E-learning recommendation framework based on deep learning. In: Book E-Learning Recommendation Framework Based on Deep Learning, pp. 455–460. IEEE (2017)
36. Hung, M.-L., Chou, C.: Students' perceptions of instructors' roles in blended and online learning environments: a comparative study. Comput. Educ. **81**, 315–325 (2015)
37. Arbaugh, J.B.: System, scholar or students? Which most influences online MBA course effectiveness? J. Comput. Assist. Learn. **30**(4), 349–362 (2014)
38. Kordrostami, M., Seitz, V.: Faculty online competence and student affective engagement in online learning. Mark. Educ. Rev. **32**(3), 240–254 (2022)
39. Shi, Y.: Course design and student learning outcomes in asynchronous online courses: The role of student motivation. The University of Texas at Austin (2022)
40. Adhikari, B.P., Koehler, T.: Understanding the impact of E-learning dimensions in advancing digital pedagogy in nepalese higher education (2024)
41. Zimmerman, B.J.: A Social cognitive view of self-regulated academic learning. J. Educ. Psychol. **81**(3), 329–339 (1989)
42. Delcker, J., Heil, J., Ifenthaler, D., Seufert, S., Spirgi, L.: First-year students AI-competence as a predictor for intended and de facto use of AI-tools for supporting learning processes in higher education. Int. J. Educ. Technol. High. Educ. **21**(1) (2024)
43. Chang, D.H., Lin, M.P.-C., Hajian, S., Wang, Q.Q.: Educational design principles of using AI chatbot that supports self-regulated learning in education: goal setting, feedback, and personalization. Sustainability **15**(12951) (2023)
44. Ouyang, F., Wu, M., Zheng, L., Zhang, L., Jiao, P.: Integration of artificial intelligence performance prediction and learning analytics to improve student learning in online engineering course. Int. J. Educ. Technol. High. Educ. **20**(1) (2023)
45. Banihashem, S.K., Kerman, N.T., Noroozi, O., Moon, J., Drachsler, H.: Feedback sources in essay writing: peer-generated or AI-generated feedback?. Int. J. Educ. Technol. High. Educ. **21**(1) (2024)
46. Basnet, R.B., Johnson, C., Doleck, T.: Dropout prediction in MOOCs using deep learning and machine learning. Educ. Inf. Technol. **27**, 11499–11513 (2022)
47. Qazdar, A., Er-Raha, B., Cherkaoui, C., Mammass, D.: A machine learning algorithm framework for predicting students performance: a case study of baccalaureate students in Morocco. Educ. Inf. Technol. **24**, 3577–3589 (2019)
48. Alsharhan, A., Salloum, S.A., Aburayya, A.: Using e-learning factors to predict student performance in the practice of precision education. J. Legal Ethical Regulatory Issues **24**(6), 1–14 (2021)
49. King, J., South, J.: Reimagining the role of technology in higher education: a supplement to the national education technology plan. In: Book Reimagining the Role of Technology in Higher Education: A Supplement to the National Education Technology Plan. U.S. Department of Education (2017)
50. Pugliese, L.: Adaptive learning systems: Surviving the storm. In: Book Adaptive learning systems: Surviving the storm (2016)

51. Qi, C., Tang, Y., Lei, Y.: Does feedback from ChatGPT help? Investigating the effect of feedback from both teacher and chatgpt on students' learning outcomes. In: Book Does Feedback from ChatGPT help? Investigating the Effect of Feedback from both Teacher and ChatGPT on Students' Learning Outcomes (2024)

Exploring the Role of Artificial Intelligence in Small and Medium Enterprises for Improved Decision-Making: A Scoping Review

Oghenetejiri Praise Ekrokpe(✉), Uchitha Jayawickrama, and Cihan Butun

Loughborough Business School, Loughborough University, Loughborough LE11 3TU, UK
{o.p.ekrokpe,u.jayawickrama,c.butun}@lboro.ac.uk

Abstract. Artificial intelligence (AI) has emerged as a transformative tool in various industries, offering significant opportunities for Small and Medium Enterprises (SMEs) to improve their decision-making processes. This study systematically explores AI applications in SMEs, highlighting trends and industry-specific implementations. Using the Preferred Reporting Items for Systematic Reviews and Meta-Analyses extension for Scoping Reviews (PRISMA-ScR), the study synthesises insights from 97 sources, including academic and non-academic literature across multiple industries. The findings show a rising adoption of AI, particularly in generative AI, machine learning, and robotic process automation, with a notable increase in publications and real-world implementations in recent years. Geographically, AI adoption in SMEs is most pronounced in developed economies such as the United States, Germany, and China, where advanced AI infrastructure, government incentives, and digital ecosystems facilitate implementation. In emerging economies, particularly India, AI is gaining traction in manufacturing and supply chain management. Sector-wise, manufacturing exhibits the highest AI integration, leveraging robotics and machine learning for automation and optimisation, followed by the finance sector, which increasingly employs AI for fraud detection and risk assessment. This study provides an overview of key AI applications in SMEs and highlights future research directions to improve AI-driven decision-making in these enterprises.

Keywords: Artificial Intelligence · SMEs · Decision Support Systems · Scoping Review · PRISMA-ScR

1 Introduction

Small and medium enterprises (SMEs) form the backbone of the global economy, comprising 90% of businesses and over 50% of employment worldwide [52]. In the

B. Delibašić et al. (Eds.): ICDSST 2025, LNBIP 546, pp. 89–109, 2025.
https://doi.org/10.1007/978-3-031-90863-7_7

European Union (EU), SMEs represent 99.8% of enterprises, contributing significantly to economic growth [16]. The European Commission defines SMEs as businesses with fewer than 250 employees and either an annual turnover not exceeding €50 million or a balance sheet total below €43 million [17]. Similarly, the U.S. Small Business Administration (SBA) classifies SMEs based on industry-specific thresholds, generally including firms with fewer than 500 employees [45].

Artificial intelligence (AI) is the simulation of human intelligence in machines, enabling tasks such as learning, reasoning, and self-correction [37]. AI applications-including machine learning (ML), predictive analytics, natural language processing (NLP), and robotic process automation (RPA)-offer SMEs opportunities to enhance efficiency, reduce costs, and optimise decision-making [39]. AI adoption has surged, with 72% of companies integrating AI into at least one business function in 2024, up from 55% in 2023 [24].

Despite increasing research on AI in SMEs, studies remain fragmented [39], lacking a systematic synthesis of AI applications across industries. Existing literature often overlooks sector-specific AI adoption trends and their impact on decision-making within SMEs. To address this gap, this study conducts a scoping review to systematically map AI applications in small businesses, examining their role in improving decision making across industries. This review follows the Preferred Reporting Items for Systematic Reviews and Meta-Analyses for Scoping Reviews (PRISMA-ScR) methodology [44], which ensures a transparent and structured approach to identifying, screening, and selecting relevant literature.

This study contributes to the literature by:

1. providing a structured analysis of AI applications in SMEs,
2. identifying key industry-specific adoption patterns, and
3. offering strategic insights for policymakers, business leaders, and researchers aiming to maximise the potential of AI in SMEs.

1.1 Objectives and Research Questions

The primary objective of this study is to systematically examine how AI applications are used in SMEs and how these applications improve decision-making in SMEs. This study is guided by the following research questions:

RQ1. What are the key AI applications across various industries in SMEs?

RQ2. How are AI applications improving decision-making in SMEs?

1.2 Methodological Rationale

Given the wide range of AI applications in SMEs, a scoping review is ideal to address research questions. Scoping reviews systematically gather, describe, and categorise relevant evidence [33]. Using the PRISMA-ScR protocol ensures transparency, reproducibility, and thorough source coverage. This method integrates findings from diverse sources, including empirical studies, conceptual frameworks, and non-academic literature. By synthesising these insights, this study improves the understanding of the transformative role of AI in SMEs and provides practical recommendations for researchers and practitioners.

2 Methods

Methods. This study used a scoping review guided by PRISMA-ScR [44] to systematically identify and map existing research on AI applications in SMEs [25]. The PRISMA-ScR protocol contains a checklist of 20 essential reporting items and two optional items: (i) Critical appraisal of individual sources of evidence and (ii) Critical appraisal within sources of evidence [44]. We excluded optional appraisals, prioritising literature mapping over methodological quality assessment. Subsequently, content-based research [30] was conducted to explore article content and descriptive analysis to identify trends and patterns in AI adoption across various SME sectors. Figure 1 describes the conceptual scheme and methodological approach of the study. The dataset identification process began with selecting five databases for comprehensive coverage of academic and non-academic literature on AI applications in SMEs. Detailed search strategies and filters are described in Sects. 2.3 and 2.4, with eligibility criteria outlined in Sect. 2.2. Following the PRISMA-ScR protocol [44], we systematically explored the literature, applying its checklist (e.g., information sources, search, data charting) to guide the process. After source selection, content-based research [30] analysed the article content to map the trends in AI adoption in SMEs.

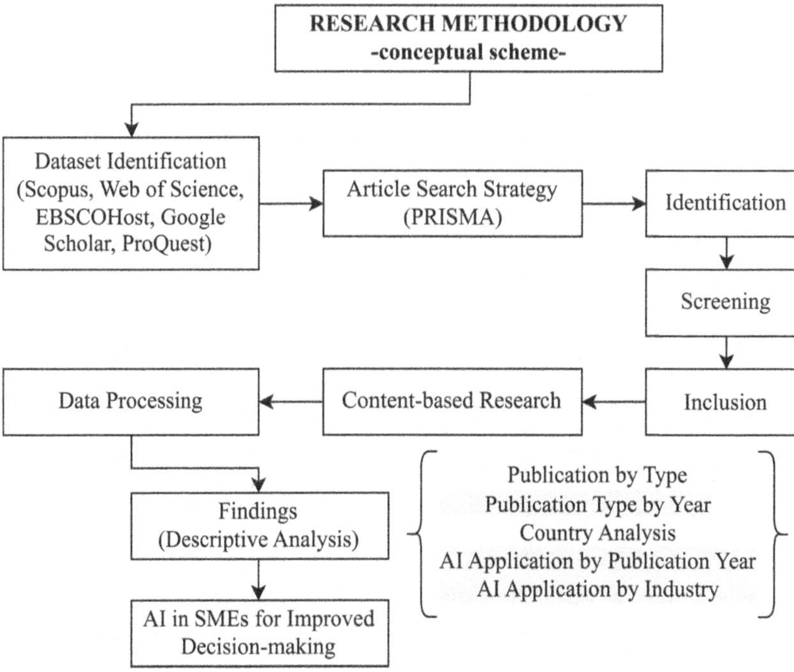

Fig. 1. Research methodology-conceptual scheme [30]

2.1 Protocol and Registration

The systematic review process for the study was guided by a protocol developed a prior to assess key AI applications across industries and their impact on decision-making in SMEs. The protocol specified objectives, eligibility criteria, data sources, search strategy, data extraction, and synthesis methods, ensuring adherence to the PRISMA-ScR guidelines for transparency and reproducibility.

2.2 Eligibility Criteria

The relevance of articles retrieved from the search process was assessed based on predefined inclusion and exclusion criteria. Studies were included if they specifically examined AI applications in SMEs, while those discussing AI broadly or focussing on large enterprises or firms that did not meet the EU or US SME definition were excluded to maintain the scope of the study. Only articles published in English were included to facilitate accurate interpretation and analysis. The review covered publications from 2010 to 2024, including early adoption, the growth phase, and current advances in AI. The 2010 starting point was selected to capture modern AI developments, including machine learning, deep learning, and big data analytics, which became increasingly accessible to SMEs [9]. This timeframe also aligns with previous studies, ensuring comparability and relevance while excluding outdated technologies.

2.3 Information Sources

A comprehensive search for academic and non-academic literature was conducted in October 2024. The search covered Scopus, Web of Science, EBSCOhost, ProQuest, and Google Scholar for academic sources and included targeted Google searches for non-academic literature using specific keywords. Academic sources comprised journal articles, conference papers, and book chapters, while non-academic sources included technical papers, research reports, industry reports, press releases, blog posts, and newspapers. To remove duplicates, the final search results-including titles, authors, and abstracts-were exported in RIS (Research Information Systems) format and managed using Zotero Reference Manager.

2.4 Search

During the search process, results were filtered to include only articles with titles containing AI- and SME-related terms [27], ensuring a more targeted and relevant selection of studies (see Appendix). Non-academic literature, including blog posts, newspapers, magazines, and press releases, was primarily retrieved from the ProQuest database using the same keywords. Additionally, targeted Google searches were conducted to identify relevant white papers and industry reports.

2.5 Selection of Sources of Evidence (Selection Process)

Following a comprehensive search across multiple databases and targeted searches for non-academic literature, the retrieved sources underwent a two-stage screening process. In the first stage, titles and abstracts were reviewed to assess their relevance to the study's scope, with ineligible articles excluded. Particular attention was given to journal quality, verified using the Scimago Journal Ranking (SJR) to eliminate articles from predatory publishers. In the second stage, the full texts of the remaining articles were thoroughly evaluated against the eligibility criteria. Access to unavailable but relevant articles was requested and granted through the Loughborough University library.

2.6 Data Charting Process

Data extraction was conducted by exporting the bibliographic details of the included articles from Zotero Reference Manager to MS Excel, facilitating systematic organisation and analysis. The charted data captured key details, including publication year, geographic focus, publication type, discussed AI technologies, and the targeted industry.

2.7 Data Items and Synthesis of Results

The bibliographic data exported to MS Excel were supplemented with additional information aligned with the review questions to support the coding process. The extracted data items include article type, publication year, industry focus, geographic scope, methodological approach, and AI application. Table 1 outlines the classification dimensions and options used for categorising the included articles, following the approach of [30].

3 Results

This section presents the results of the literature search, detailing the sources of evidence screened, assessed for eligibility, and included in the review, as illustrated in Fig. 2. In addition, charts and graphs aligned with the review objectives are provided to summarise key insights from the selected sources.

3.1 Selection of Sources of Evidence (Results of Literature Search)

Database searches yielded 1,174 records, distributed across Scopus (n = 478), ProQuest (n = 223), EBSCOhost (n = 257), Google Scholar (n = 134) and Web of Science (n = 82). In addition, targeted Google searches identified 22 more records. During the identification stage, 425 duplicate records and 11 retracted articles were removed, leaving 738 records for further evaluation. These records were screened for relevance based on their titles and abstracts, resulting in the exclusion of 552 records due to: (i) a primary focus on challenges

Table 1. Framework for content-based analysis.

Investigated Aspect	Dimension	Option	Option Definition
Methodology	Approach	Conceptual/empirical	The paper provides a conceptual framework/model, or empirical applications (e.g., surveys, case studies, or experiments.
Geography	The location	Country/region/global	The paper focuses on a single country, a region, or a global perspective.
Industry	Industry focus	Specific industry/generic	The paper is set in a specific industry (e.g., healthcare, manufacturing) or discusses generic SMEs without industry-specific focus.
AI Application	AI technology adopted	Specific application/various /generic	The paper examines specific AI applications (e.g., predictive analytics, chatbots) or general AI adoption in SMEs.

and barriers to AI adoption, (ii) publication in predatory journals or by questionable publishers, and (iii) false positives, such as papers using "SMES" for "superconducting magnetic energy storage" or containing terms like "smell," "smelting," or "smear."

Following screening, 186 full-text articles were retrieved and assessed for eligibility. At this stage, 95 articles were excluded due to: (i) lack of direct discussion on AI applications in SMEs (n = 58), (ii) journal quality (n = 28), (iii) thesis publications (n = 3), (iv) articles withdrawn by the author or publisher (n = 1), and (v) unreliable newspaper articles (n = 5). Of the 22 records obtained from Google searches, nine were excluded for discussing AI in a general sense or outside the SME context, and seven for lacking a focus on AI applications in SMEs. The final selection process resulted in the inclusion of 97 studies.

3.2 Characteristics of Sources of Evidence

A total of 97 studies were included in the review, comprising 64 academic articles and 33 non-academic sources. These sources examined AI applications in SMEs across various sectors, including finance, manufacturing, and healthcare. The studies had a global scope, with most focusing on North America, Europe, and Asia. The methodologies employed ranged from conceptual frameworks to empirical studies, including surveys, case studies, and interviews. Empirical studies dominated the literature, underscoring real-world applications of AI. The data

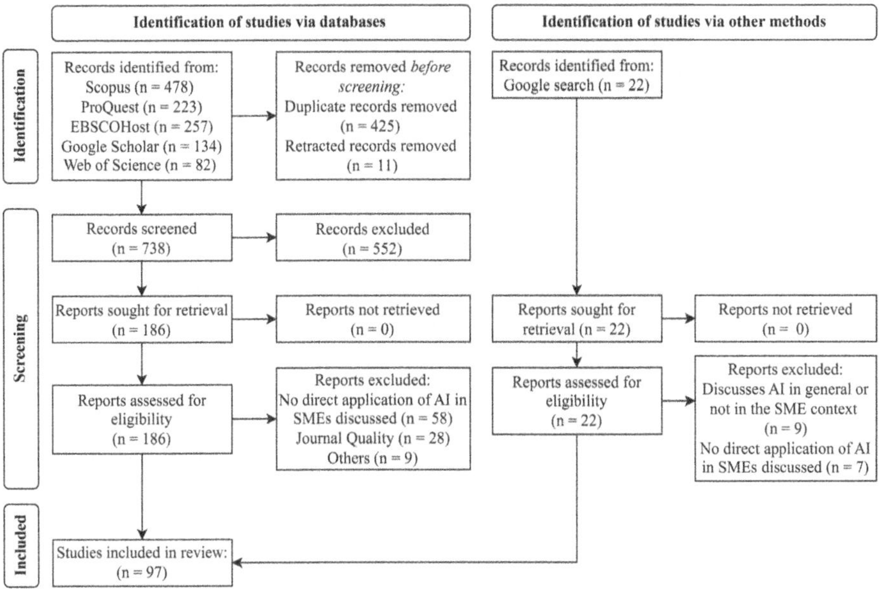

Fig. 2. PRISMA flow diagram [28]

charted from these 97 sources provided a comprehensive overview of AI adoption in SMEs across different industries.

3.3 Results of Individual Sources of Evidence

The detailed data charted from the sources of evidence are available in a supplementary file (accessible here: https://bit.ly/individualsourcesofevidence). Given the extensive number of sources reviewed, the information has been organised into a structured spreadsheet. For each included source, key characteristics are reported, including author(s), title, publication year, and type, methodological approach, geographic context, industry/sector, AI applications, and link/DOI.

3.4 Synthesis of Results

Publication Types and Trends. The sources of evidence included in this review cover a diverse range of publication types, as illustrated in Fig. 3. A significant portion of the evidence comes from journal articles (40), while non-academic contributions include press releases (17), newspaper articles (5), and magazine articles (4), among others. The temporal distribution of publication is represented in Fig. 4, which highlights the increased research interest in AI applications in SMEs. The stacked column chart displays the total number of publications within each category per year, with each stack divided into segments representing different types of publication. The numbers indicate the count of publications in each category for the corresponding year. During the early period

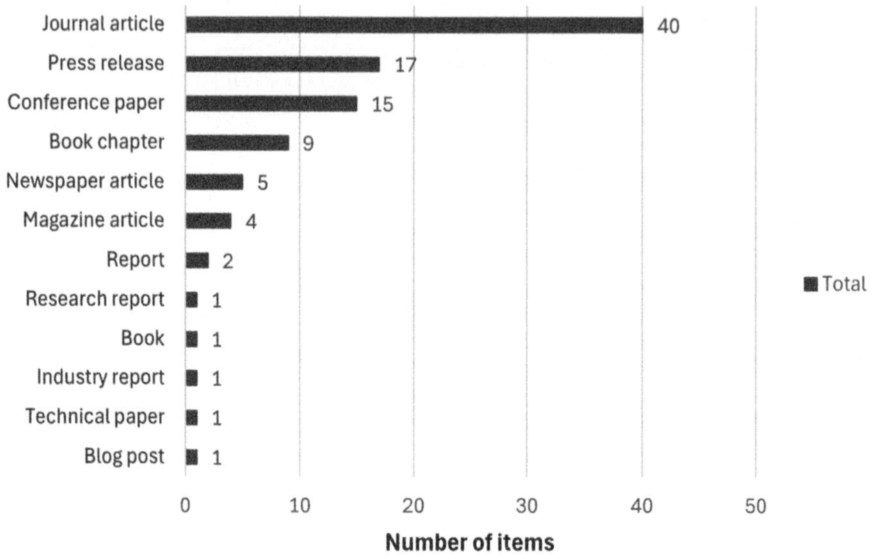

Fig. 3. Summary of various types of publications

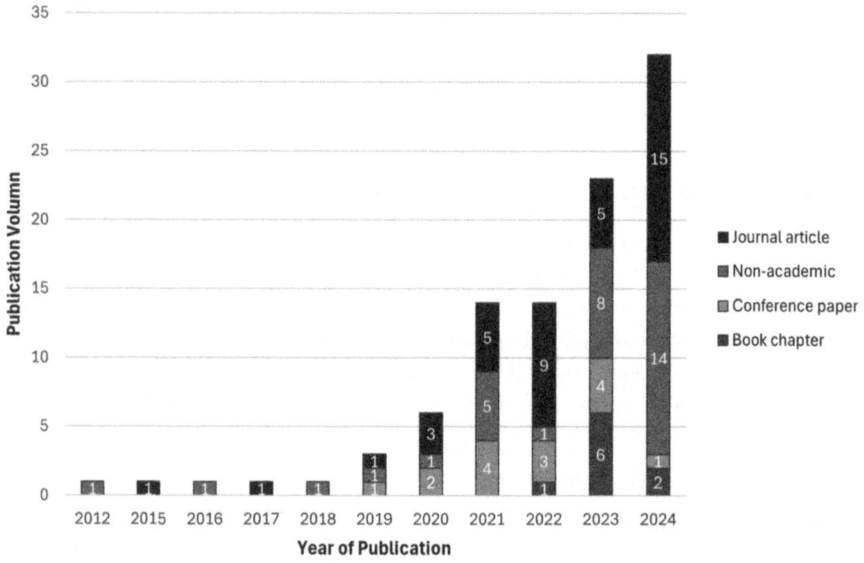

Fig. 4. Publication trends

(2012-2018), only a limited number of publications were observed, mainly journal and conference articles, reflecting an emerging interest in the field. From 2019 to 2021, the volume of publications increased steadily, with notable contributions from conference papers and journal articles. Most publications (71.1%)

have appeared in the last three years, demonstrating the expanding role of AI in SMEs. In particular, 2023 saw an increase in both academic and non-academic sources (23.7%), signaling the accelerating adoption of AI across various sectors. In 2024, the highest number of publications (32.9%) was recorded, with journal articles (15.5%) and non-academic sources (14.4%) contributing significantly, underscoring the growing relevance of AI and its practical implementation in SMEs.

Global Distribution of AI Applications in SMEs. The evidence sources analysed in this study demonstrate the global reach of AI applications in SMEs across various industries, as shown in Fig 5. The US (18 instances) leads, reflecting its robust AI ecosystem in sectors such as manufacturing [48] and finance [8]. Germany (7) stands out for its leadership in advanced manufacturing with the widespread use of AI in industrial automation [50] and energy optimization [51]. China (6) follows closely, with AI applications in accounting [55], e-commerce [49], and healthcare [15]. The United Kingdom (6) on AI in cybersecurity [35], and RPA in the food industry [32]. India (6) use AI to automate manufacturing processes [43] and workforce management [23]. The United Kingdom (6) focusses on AI in cybersecurity [35] and RPA in the food industry [32], while India (6) applies AI to automate manufacturing processes [43] and workforce management [23]. Denmark (4) is notable for its use of AI in manufacturing [19], particularly in collaborative robotics (cobots) [36] and predictive analytics [10]. Spain (3) uses AI in industrial robotics [6] and cybersecurity [13], while Saudi Arabia (3) highlights the role of AI in B2B [5] and social media marketing [7]. Other countries, such as Taiwan, Malaysia, Canada, Japan, Czech Republic, Fin-

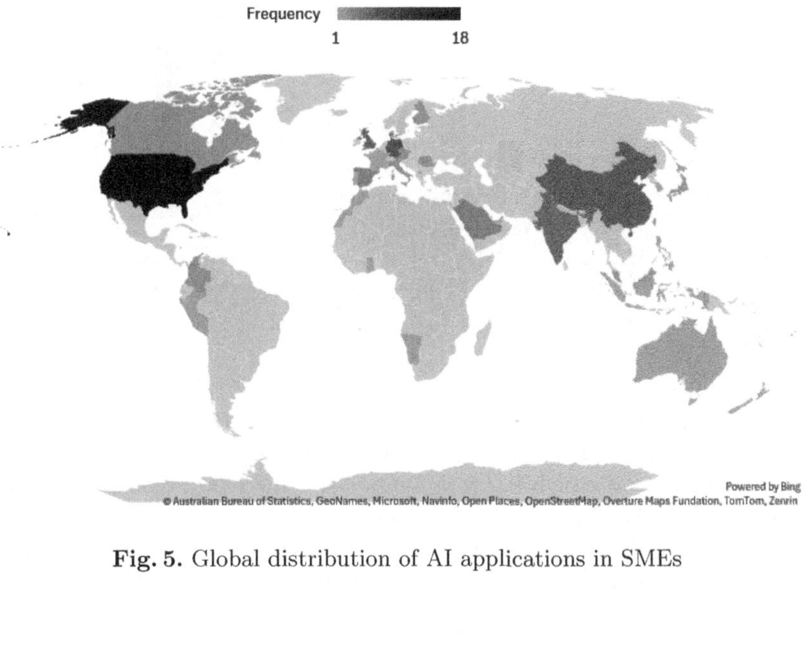

Fig. 5. Global distribution of AI applications in SMEs

land and Romania (2 each), and Italy, France, Indonesia, South Korea and North Macedonia (1 each), showcase AI-focused applications in areas such as food production [47], healthcare [21], manufacturing [29], and finance [12]. In addition, countries such as Ghana, Portugal, Morocco, Lebanon, and Peru contribute to the diversity of AI applications, particularly in marketing [1] and retail [11].

Adoption of AI in SMEs Over Time. Figure 6 illustrates the distribution of AI applications in SMEs over time. The numbers on the stacked column chart represent the total count of AI applications identified in publications for each year. Each stack is divided into segments that correspond to different AI applications.

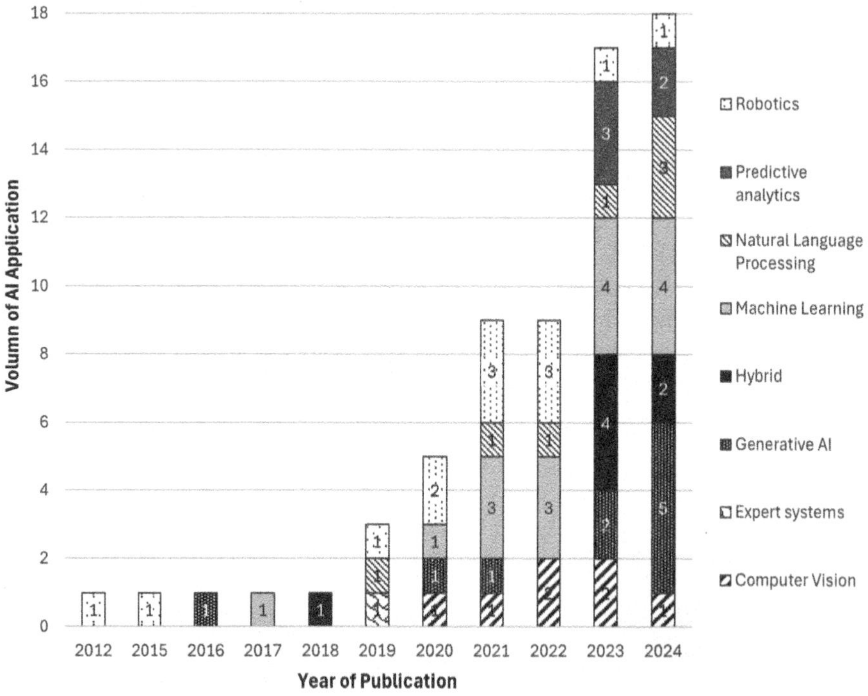

Fig. 6. Trends in AI adoption in SMEs over time

AI Applications in SMEs Across Industries and Sectors. The stacked bar chart in Fig. 7 presents the distribution of AI applications across various industries and sectors. Each bar represents a specific sector or industry, and the segments within the bar correspond to different AI applications. The length of each segment indicates the number of publications featuring that specific AI application within the sector or industry. Manufacturing is the leading sector in AI adoption, with the highest number of AI applications. Robotics (9) and machine learning (8) are the most widely used technologies in this sector, followed by computer vision (6).

The finance sector has adopted various AI technologies, including predictive analytics, hybrid applications, machine learning, and expert systems. Marketing and advertising predominantly leverage generative AI (GenAI). Emerging AI adoption in sectors such as cybersecurity, food, HR services, retail, and medical technology (MedTech) is visible, each integrating a small but diverse set of AI technologies. However, there is limited AI deployment in specialised sectors such as telecommunications, web services, agriculture, banking, financial technology (Fintech), and insurance, with each industry reporting only a single AI application in different categories.

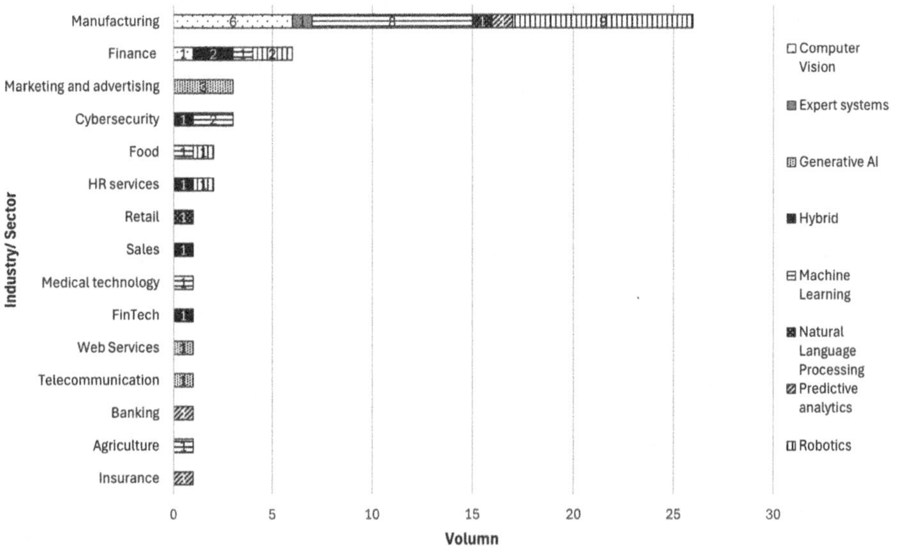

Fig. 7. AI adoption in SMEs across various industries and sectors

4 Discussion

In this scoping review, we identified 97 studies from academic and non-academic sources addressing the applications of AI in SMEs across various industries, published between 2010 and 2024. Our findings indicate a growing interest in this area, with the last three years accounting for 71.1% of publications, and 2023 and 2024 showing the highest numbers (23.7% and 32.9%, respectively). This surge reflects the increasing recognition of AI to improve decision-making and competitiveness in SMEs.

Deepthi and Bansai [14] noted that while the first publication on AI applications in SMEs dates back to 1989, significant research activity only began to accelerate around 2019, with the majority of studies emerging in the last few years, particularly between 2019 and 2021. This suggests that AI adoption in SMEs has gained substantial academic and practical attention more recently. Similarly, Schwaeke et al. [39] describe the literature on AI adoption in SMEs as fragmented and lacking coherence in previous years.

The increased focus on AI adoption can be attributed to the recognition of SMEs as key drivers of economic growth, prompting investments in training curricula to support their digital transformation [39]. Additionally, the broader shift toward the Fourth Industrial Revolution (Industry 4.0), characterised by digitalisation and automation, has significantly influenced AI adoption. SMEs are increasingly relying on AI-driven real-time data analysis, predictive analytics, and automation to improve productivity, reduce errors, and lower operational costs [14].

The adoption of AI in SMEs exhibits a global distribution, with the US emerging as the leading contributor (18 instances), highlighting the country's significant role in driving AI research and implementation within SMEs [14]. This is partly due to the robust support provided by the US Small Business Administration (SBA) [46] and related initiatives, which offers extensive financial and non-financial assistance to SMEs, including de-risked loans for acquiring essential assets, as well as personalised business counselling [26].

In addition, the launch of the Small Business Digital Alliance (SBDA) in 2022, in partnership with Business Forward, Inc. [41], played a crucial role in accelerating digital transformation among SMEs. The SBDA provides critical resources, helping small businesses integrate digital tools such as e-commerce platforms, data analytics, and AI-driven marketing solutions to enhance their scalability and competitiveness [26]. The initiative's Digital Tools Library [40] further supports SMEs by offering free resources from leading US tech companies, equipping them with the necessary knowledge and skills to effectively leverage AI and other advanced technologies. Furthermore, the SBA fosters collaboration between entrepreneurs, policymakers, and industry leaders, facilitating connections that support SME growth in the digital economy. These concerted efforts, coupled with the broader technological ecosystem and funding opportunities available in the US, have placed the country at the forefront of AI adoption in SMEs.

Sector-wise, manufacturing leads in AI adoption, with robotics (9) and machine learning (8) being the most widely used technologies. Robotics plays a crucial role in this transformation, as cobots are designed to work alongside human operators-a key principle of the Fifth Industrial Revolution (Industry 5.0) to enhance efficiency, customisation, and sustainability in manufacturing environments [34]. These robots are particularly beneficial for SMEs, which often operate in dynamic settings with varying product demands.

The integration of robotics allows for improved production throughput, while reducing labour costs and improving workplace safety [53]. Machine learning complements these advancements by enabling data-driven decision-making. It facilitates the analysis of large datasets to identify patterns and optimise processes, thereby improving operational efficiency. For example, machine learning algorithms can predict equipment failures, optimise supply chain logistics, and improve quality control measures, which are critical elements in maintaining competitiveness in the manufacturing sector [50].

In the finance sector, AI applications are notably diverse, encompassing areas such as risk management and fraud detection [22]. AI technologies, particularly machine learning, analyse vast amounts of transaction data to detect patterns indicative of fraud. These systems can learn from historical data to identify anomalies in real-time, allowing financial institutions to issue timely alerts and prevent fraudulent activities. Schwaeke et al. [39] emphasise that AI-driven financial analytics allow SMEs to make more informed investment and budgeting decisions. AI-powered platforms, such as Intuit Assist [20], provide personalised financial recommendations, reducing the complexity of financial planning and supporting faster, data-driven decision-making.

Robotic Process Automation (RPA) in bookkeeping [4] streamlines financial processes by automating repetitive tasks such as invoice processing, data entry, and financial reporting. AI technologies such as Optical Character Recognition (OCR) and machine learning facilitate accurate journal entries, reducing manual data entry errors and transforming bookkeeping into a decision-support tool. Artificial intelligence in marketing also plays an important role in SME decision-making, particularly in customer behaviour analysis and predictive analytics. Abrokwah-Larbi and Awuku-Larbi [1] note that AI-driven marketing tools enable SMEs to improve decision quality by making accurate predictions about consumer behaviour, overcoming cognitive limitations. This improves SMEs' ability to personalise marketing strategies, optimise advertising efforts, and improve overall business performance.

Several case studies illustrate the role of AI in improving decision-making within SMEs. Fuentes et al. [18] presented and validated ACODAT (Autonomous Cycles of Data Analysis Tasks), an AI-driven architecture for automating production chains in the agro-industry. In a case study, a coffee producing SME utilised ACODAT to analyse inputs and environmental variables such as temperature and humidity, to make cost decisions and optimise resources. In Germany, a metalworking SME implemented a movable welding cobot, enabling welders to decide which tasks to automate, which improved productivity by shifting repetitive tasks to the cobot [38]. Similarly, a Danish glass manufacturing SME used simulations to optimise a cobot work cell, increasing production from 15 to 45 parts daily by informing layout and deployment decisions [36].

In bookkeeping, a Finnish event management SME adopted Procountor to automate financial reporting, eliminating tedious manual submission processes. This enabled real-time invoice processing, reduced administrative workload, and allowed the company to focus on core business activities [4]. Likewise, a US beverage distributor in Alabama and North Carolina utilised Beanworks for invoice management, enhancing visibility and speeding approval decisions remotely, increasing efficiency during demand spikes [4].

The results also indicate that the adoption of GenAI increased significantly in 2023 and 2024, particularly in the marketing and advertising sector, highlighting the growing use of AI-generated content for personalised marketing and advertising. A recent survey conducted by the Organisation for Economic Co-operation and Development (OECD) under the OECD Digital for SMEs (D4SME) Global Initiative, covering seven OECD countries (France, Germany, Italy, Japan, Korea, Spain, and the US), highlights key trends in SME adoption of digital tools, particularly AI applications. The findings indicate that SMEs have rapidly embraced GenAI, with nearly 18% of respondents adopting such tools within a year of large language models (LLMs) becoming publicly available in 2022 [26]. However, only 25% of surveyed SMEs actively integrate AI into their core operations, either by directly leveraging GenAI services, developing custom AI systems, or passively using AI-embedded platforms. Since the survey sample includes only SMEs already engaged with large digital platforms, many of which embed machine learning algorithms, the findings suggest that all respondents are at least indirectly utilising AI in some capacity. This underscores the growing ubiquity of AI in SME operations, even among businesses that may not actively develop AI-driven [26].

Several companies are leveraging GenAI to improve various aspects of SME operations, from communication and marketing to business planning and customer engagement. Aircall [2] has integrated AI-driven transcription into its communication platform, enabling sales and support teams to transcribe calls and voicemails efficiently. This automation reduces manual effort, enhances customer interactions, and provides valuable insights for performance monitoring and team training. Similarly, GoDaddy [3] employs GenAI in its suite of products, particularly through GoDaddy Studio, which generates professional design templates for websites and social media. This allows SMEs to establish an online presence without requiring specialised design skills.

In digital advertising, Pencil [31] offers an AI-powered platform that generates optimised video ads by analysing historical ad performance, enabling small businesses to create high-quality marketing content. Meanwhile, Speedy-Brand [42] specialises in AI-driven content marketing by offering search engine optimised (SEO) blog and social media content, helping businesses strengthen their digital presence while reducing the costs and time associated with content creation. For business planning and operations, ZenBusiness [54] provides AI-driven tools that simplify business planning through guided question-based automation, making essential business tasks more accessible to entrepreneurs without technical expertise.

4.1 Limitations

Despite offering valuable insights into the adoption of AI in SMEs and how it improves decision-making, this study has several limitations. First, the review primarily relies on published academic and non-academic sources, which may introduce publication bias, as studies reporting positive AI outcomes are more likely to be published. In addition, the majority of sources analysed focus on developed economies, particularly the US and EU, potentially overlooking AI adoption trends in emerging markets where SMEs face different challenges, such as infrastructure constraints, regulatory barriers, and limited access to digital resources. Another limitation lies in the lack of empirical validation, as the study does not include primary data collection from SMEs. Without direct insights from SME owners, managers, or employees, the findings may not fully reflect the practical benefits and challenges experienced by businesses implementing AI solutions. Finally, while the study highlights the role of AI in decision-making, it does not extensively explore the long-term business impacts, including financial performance, workforce dynamics, and sustainability concerns.

4.2 Future Work

To address these limitations, future research should adopt a more comprehensive and empirical approach to studying AI adoption in SMEs. First, conducting primary research through surveys, interviews, or case studies with SME owners and managers would provide deeper insights into real-world implementations and success factors. Comparative studies across different economic regions, including emerging markets, would provide valuable information on how AI adoption varies based on local regulatory environments, digital infrastructure, and financial accessibility. Furthermore, future studies should explore the ethical and regulatory challenges associated with AI adoption in SMEs, particularly concerning data privacy, algorithmic biases, and compliance with evolving legal frameworks. Given the rapid advancements in AI, longitudinal studies tracking the impact of AI adoption over time would be beneficial in understanding how AI influences SME growth, productivity, and workforce adaptation in the long run. Finally, interdisciplinary research incorporating perspectives from business, technology, and policy-making could offer holistic strategies for SMEs to maximise AI. Investigating AI-driven business models, workforce reskilling initiatives, and strategies to integrate AI into SME ecosystems could provide actionable insights for policymakers, industry leaders, and SME practitioners aiming to leverage AI effectively in their operations.

4.3 Conclusion

This study aimed to explore the role of AI in SMEs for improved decision-making by addressing two research questions. A comprehensive search for relevant academic and non-academic literature was conducted across five databases, supplemented by targeted Google searches for additional non-academic sources, following the PRISMA-ScR framework. After initially identifying 1,196 records, a rigorous screening and eligibility assessment process yielded 97 sources of evidence for full review and analysis. We utilised content-based research and descriptive analysis to derive key findings related to the research questions. The findings indicate that the adoption of AI in SMEs has gained significant traction in recent years, particularly since 2019, driven by the broader transition to Industry 5.0 and the increasing availability of AI technologies designed for SMEs. The analysis reveals that AI is increasingly used in finance and manufacturing to optimise business operations, improve efficiency, and support strategic decision-making. In particular, the integration of generative AI and other AI-driven solutions, such as automated bookkeeping, financial analysis, and intelligent marketing strategies, has substantially improved decision-making capabilities within SMEs. Furthermore, this study underscores the pivotal role of government support in facilitating AI adoption, with evidence from successful US initiatives suggesting that similar initiatives could be adopted in other regions to accelerate SMEs' digital transformation.

Acknowledgments. This study was funded by Loughborough University. The support enabled access to essential research materials and resources for the successful completion of this study.

Appendix

(TITLE ((“Artificial intelligence” OR “AI” OR “machine learning” OR “deep learning” OR “neural network* ” OR “automation” OR “intelligent system* ” OR “cognitive comput* ” OR “expert system* ” OR “predictive analy* ” OR “natural language processing” OR “NLP* ” OR “robot* ” OR “robotic process automation” OR “RPA* ”)) AND TITLE ((“SME* ” OR “SMB*” OR “Small and Midsize Business* ” OR “small and medium enterpris* ” OR “small and medium-sized enterpris* ” OR “small and medium business* ” OR “small and medium-sized business* ” OR “small and medium compan* ” OR “small and medium-sized compan* ” OR “micro small and medium enterprise* ” OR “micro small and medium-sized enterprise* ” OR “MSME* ” OR “micro enterpris* ” OR “micro-sized enterpris* ” OR “micro compan* ” OR “micro business* ” OR “micro-sized business* ” OR “micro firm* ” OR “micro-sized firm* ” OR “small enterpris* ” OR “small-sized enterpris* ” OR “small compan* ” OR “small-sized compan* ” OR “small business* ” OR “small sized-business* ” OR “small firm* ” OR “small-sized firm* ” OR “medium enterpris* ” OR “medium-sized enterpris* ” OR “medium compan* ” OR “medium-sized compan* ” OR “medium business* ” OR “medium-sized business* ” OR “medium firm* ” OR “medium-sized firm* ”)) AND ALL((“adopti* ” OR “practic* ” OR “application* ” OR “implement* ” OR “integrat* ” OR “use” OR “utili?ation* ”)) AND ALL((“industr* ” OR “sector* ”)) AND ALL((“impact” OR “effect*” OR “benefit* ” OR “success* ” OR “outcome* ” OR “performance” OR “efficiency” OR “innovation* ” OR “productivity”)) AND PUBYEAR > 2009 AND PUBYEAR < 2025 AND (LIMIT-TO (LANGUAGE, “English”)).

References

1. Abrokwah-Larbi, K., Awuku-Larbi, Y.: The impact of artificial intelligence in marketing on the performance of business organizations: evidence from smes in an emerging economy. J. Entrepreneurship Emerg. Econom. **16**, 1090–1117 (2024). https://doi.org/10.1108/JEEE-07-2022-0207
2. Aircall: Aircall launches new AI features that empower SMBs to leverage daily conversations to drive performance (2023). https://www.businesswire.com/news/home/20230405005223/en/Aircall-Launches-New-AI-Features-That-Empower-SMBs-to-Leverage-%Daily-Conversations-to-Drive-Performance. Accessed 13 Jan 2025
3. Ashish Pandey & The Economic Times: Ai opens up a world of possibilities for smes: Godaddy (2024). https://economictimes.indiatimes.com/small-biz/sme-sector/ai-opens-up-a-world-of-possibilities-for-smes-godaddy/articleshow/114500109.cms?from=mdrhttps://www.godaddy.com/en-uk?msockid=2f7748c783786ad30df25d8d82ba6b64
4. Azman, N.A., Mohamed, A., Jamil, A.M.: Artificial intelligence in automated bookkeeping: a value-added function for small and medium enterprises. JOIV : Int. J. Inform. Visualization **5**, 224 (2021). https://doi.org/10.30630/joiv.5.3.669
5. Baabdullah, A.M., Alalwan, A.A., Slade, E.L., Raman, R., Khatatneh, K.F.: SMEs and artificial intelligence (AI): antecedents and consequences of AI-based B2B practices. Ind. Mark. Manage. **98**, 255–270 (2021). https://doi.org/10.1016/j.indmarman.2021.09.003
6. Ballestar, M.T., Ángel Díaz-Chao, Sainz, J., Torrent-Sellens, J.: Knowledge, robots and productivity in SMEs: explaining the second digital wave. J. Business Res. **108**, 119–131 (2020). https://doi.org/10.1016/j.jbusres.2019.11.017
7. Basri, W.: Examining the impact of artificial intelligence (AI)-assisted social media marketing on the performance of small and medium enterprises: Toward effective business management in the Saudi Arabian context. Int. J. Comput. Intell. Syst. **13**, 142–152 (2020). https://doi.org/10.2991/ijcis.d.200127.002
8. Bench: Bench raises $18 million to simplify small business accounting, cuts prices through automation (2018). https://www.businesswire.com/news/home/20180131005373/en/Bench-Raises-18-Million-to-Simplify-Small-Business-Accounting-Cuts-Prices-Through-Automation. Accessed 13 Jan 2025
9. Bianchini, M., Michalkova, V.: Data analytics in SMEs: Trends and policies. OECD SME and Entrepreneurship Papers (15) (2019). https://doi.org/10.1787/1de6c6a7-en
10. Bøgh, S., Hain, D.S., Hansen, E.B., Jensen, S.B., Tvedebrink, T., Jurowetzki, R.: Predictive Analytics Applications for Small and Medium-Sized Enterprises (SMEs)—A Mini Survey and Real-World Use Cases, pp. 263–279. Springer International Publishing, Cham (2023). https://doi.org/10.1007/978-3-031-15428-7_22
11. Caycho, H.E.C., Gamboa, E.N.E., Cruz, M.A.E.: Intelligent system in micro and small business sales management: A systematic review. In: Proceedings of the 20th LACCEI International Multi-Conference for Engineering, Education and Technology: "Education, Research and Leadership in Post-pandemic Engineering: Resilient, Inclusive and Sustainable Actions". Latin American and Caribbean Consortium of Engineering Institutions (2022). https://doi.org/10.18687/LACCEI2022.1.1.287

12. Cubric, M., Li, F.: Bridging the 'concept-product' gap in new product development: emerging insights from the application of artificial intelligence in fintech smes. Technovation **134**, 103017 (2024). https://doi.org/10.1016/j.technovation.2024.103017
13. Cyberwrite: Mapfre and cyberwrite join forces to promote cyber protection for smes with AI (2023). https://www.businesswire.com/news/home/20231129755647/en/MAPFRE-and-Cyberwrite-Join-Forces-to-Promote-Cyber-Protection-for-SMEs-with-AI/. Accessed 13 Jan 2025
14. Deepthi, B., Bansal, V.: Adoption of Artificial Intelligence in Small and Medium-sized Enterprises: A Systematic Literature Review and Bibliometric Analysis of Global Research Trends, pp. 321–335. Productivity Press (2023). https://doi.org/10.4324/9781003358411-20
15. Dumbach, P., Liu, R., Jalowski, M., Eskofier, B.M.: The adoption of artificial intelligence in SMEs-a cross-national comparison in German and Chinese healthcare. In: BIR Workshops, pp. 84–98 (2021)
16. European Commission: Number of small and medium-sized enterprises (SMEs) in the Eduropean union from 2008 to 2024, by number of enterprises [graph] (2024). https://www.statista.com/statistics/878412/number-of-smes-in-europe-by-size/. Accessed 3 Feb 2025
17. European Commission and Directorate-General for Internal Market, Industry, Entrepreneurship and SMEs: User guide to the SME definition. Publications Office (2020). https://doi.org/10.2873/255862
18. Fuentes, J., Aguilar, J., Montoya, E.: Ángel Pinto: autonomous cycles of data analysis tasks for the automation of the production chain of MSMEs for the agroindustrial sector. Information **15**, 86 (2024). https://doi.org/10.3390/info15020086
19. Hansen, E.B., Iftikhar, N., Bøgh, S.: Concept of easy-to-use versatile artificial intelligence in industrial small & medium-sized enterprises. Procedia Manufact. **51**, 1146–1152 (2020). https://doi.org/10.1016/j.promfg.2020.10.161
20. Intuit Incorporated: Intuit launches ai-powered intuit assist for quickbooks, giving millions of businesses a competitive edge (2024). https://www.businesswire.com/news/home/20241120869309/en/Intuit-Launches-AI-Powered-Intuit-Assist-for-QuickBooks-Giving-Millions-of-Businesses-a-Competitive-Edge Accessed 13 Jan 2025
21. Jafarzadeh, P., Vähämäki, T., Nevalainen, P., Tuomisto, A., Heikkonen, J.: Supporting SME companies in mapping out AI potential: a finnish AI development case. J. Technol. Transf. (2024). https://doi.org/10.1007/s10961-024-10122-5
22. Kang, C.C., Zhang, L.J., Wang, J.M., Lin, C.H.: Application of Open AI in Small and Medium-Sized Enterprise (SME) Risk Assessment and Financial Analysis System, pp. 331–340. Springer Nature Switzerland, Cham (2024). https://doi.org/10.1007/978-3-031-63269-3_25
23. Kumar, M., Raut, R.D., Mangla, S.K., Ferraris, A., Choubey, V.K.: The adoption of artificial intelligence powered workforce management for effective revenue growth of micro, small, and medium scale enterprises (msmes). Production Planning Control **35**, 1639–1655 (2024). https://doi.org/10.1080/09537287.2022.2131620
24. McKinsey & Company: Adoption of artificial intelligence among organizations worldwide from 2017 to 2024, by type [graph] (2024). https://www.statista.com/statistics/1545783/ai-adoption-among-organizations-worldwide/. Accessed 03 Feb 2025
25. Mwangakala, H.A., et al.: Emerging digital technologies potential in promoting equitable agricultural supply chain: a scoping review. Technol. Forecast. Soc. Chang. **208**, 123630 (2024). https://doi.org/10.1016/j.techfore.2024.123630

26. OECD: SME Digitalisation to manage shocks and transitions: 2024 OECD D4SME survey, vol. 62. OECD Publishing (2024). https://doi.org/10.1787/eb4ec9ac-en
27. Oldemeyer, L., Jede, A., Teuteberg, F.: Investigation of artificial intelligence in SMEs: a systematic review of the state of the art and the main implementation challenges. Manage. Rev. Quart. (2024). https://doi.org/10.1007/s11301-024-00405-4
28. Page, M.J., et al.: The PRISM 2020 statement: an updated guideline for reporting systematic reviews. bmj **372** (2021). https://doi.org/10.1136/bmj.n71
29. Park, B., Kim, T., Jong-pil, J.: AI-smart factory: Design and verification of Korean SME AI smart factory using level diagnosis system. Procedia Comput. Sci. **224**, 169–179 (2023). https://doi.org/10.1016/j.procs.2023.09.025
30. Parmentola, A., Petrillo, A., Tutore, I., De Felice, F.: Is blockchain able to enhance environmental sustainability? a systematic review and research agenda from the perspective of sustainable development goals (sdgs). Bus. Strateg. Environ. **31**(1), 194–217 (2022). https://doi.org/10.1002/bse.2882
31. Pencil: Small businesses and ecommerce startups use pencil's new creative AI video ad system to level the playing field (2021). https://www.businesswire.com/news/home/20210520005364/en/Small-Businesses-and-eCommerce-Startups-Use-Pencil%E2%80%99s-New-Creative-AI-Video-Ad-System-to-Level-the-Playing-Field. Accessed 13 Jan 2025
32. Pendrous, R.: New easy-to-use Danish robot targets small food firms (2012). https://www.foodmanufacture.co.uk/Article/2012/06/07/New-robots-targeted-at-small-scale-food-firms/. Accessed 13 Jan 2025
33. Pollock, D., et al.: 'how-to:'scoping review? J. Clin. Epidemiol. 111572 (2024). https://doi.org/10.1016/j.jclinepi.2024.111572
34. Raffik, R., Sathya, R.R., Vaishali, V., Balavedhaa, S., Lakshmi, N.J.: Industry 5.0: Enhancing human-robot collaboration through collaborative robots - a review. In: 2023 2nd International Conference on Advancements in Electrical, Electronics, Communication, Computing and Automation (ICAECA), pp. 1–6 (2023). https://doi.org/10.1109/ICAECA56562.2023.10201120
35. Rawindaran, N., Jayal, A., Prakash, E., Hewage, C.: Cost benefits of using machine learning features in NIDs for cyber security in UK small medium enterprises (SME). Future Internet **13**, 186 (2021). https://doi.org/10.3390/fi13080186
36. Raza, M., Malik, A.A., Bilberg, A.: PDCA integrated simulations enable effective deployment of collaborative robots: case of a manufacturing SME. Procedia CIRP **104**, 1518–1522 (2021). https://doi.org/10.1016/j.procir.2021.11.256
37. Russell, S.J., Norvig, P.: Prentice Hall Series In Artificial Intelligence. Prentice Hall Englewood Cliffs, NJ (1995)
38. Schumacher, S., Hall, R., Waldman-Brown, A., Sanneman, L.: Technology adoption of collaborative robots for welding in small and medium-sized enterprises: A case study analysis. In: Proceedings of the Conference on Production Systems and Logistics: CPSL 2022, pp. 462–471. Hannover: publish-Ing. (2022). https://doi.org/10.15488/12176
39. Schwaeke, J., Peters, A., Kanbach, D.K., Kraus, S., Jones, P.: The new normal: The status quo of AI adoption in SMEs. J. Small Business Manage. 1–35 (2024). https://doi.org/10.1080/00472778.2024.2379999
40. Small Business Digital Alliance: Digital tool library (2022). https://www.smallbusinessdigitalalliance.com/digital-tool-library/. Accessed 02 Feb 2025

41. Small Business Digital Alliance: State of small business (2022). https://www.smallbusinessdigitalalliance.com/wp-content/uploads/2022/09/SBDA_STATEOFSMALLBUSINESS.pdf#:~:text=Launched%20in%20February%202022%2C%20the%20Small%20Business%20Digital,in%20support%20of%20policies%20promoting%20America%E2%80%99s%20economic%20comp. Accessed 02 Feb 2025
42. SpeedyBrand: Speedybrand secures $2.5m in funding led by google ventures and y combinator, introduces its AI-driven SEO content marketing solution for SMBs (2023). https://www.businesswire.com/news/home/20230713281212/en/SpeedyBrand-Secures-2.5M-in-Funding-Led-by-Google-Ventures-and-Y-Combinator-Introduces-Its-AI-Driven-SEO-Content-Marketing-Solution-for-SMBs. Accessed 13 Jan 2025
43. The Economic Times: AI visual inspection company, lincode, raises investment from accel, others (2022). https://economictimes.indiatimes.com/small-biz/sme-sector/ai-visual-inspection-company-lincode-raises-investment-from-accel-others/articleshow/88830237.cms?from=mdr, https://lincode.ai/
44. Tricco, A.C., et al.: Prisma extension for scoping reviews (PRISMA-SCR): checklist and explanation. Ann. Intern. Med. **169**(7), 467–473 (2018). https://doi.org/10.7326/M18-0850
45. U.S. Small Business Administration: Table of small business size standards (2021). https://www.sba.gov/document/support-table-size-standards. Accessed 16 Jan 2025
46. U.S. Small Business Administration: Homepage (2025). https://www.sba.gov/. Accessed 02 Feb 2025
47. Ushada, M., Okayama, T., Khuriyati, N., Suyantohadi, A.: Affective temperature control in food SMEs using artificial neural network. Appl. Artif. Intell. **31**, 555–567 (2017). https://doi.org/10.1080/08839514.2017.1390327
48. Waldman-Brown, A.: Redeployment or Robocalypse? workers and automation in Ohio manufacturing SMEs. Camb. J. Reg. Econ. Soc. **13**, 99–115 (2020). https://doi.org/10.1093/cjres/rsz027
49. Wei, R., Pardo, C.: Artificial intelligence and SMEs: How can B2B SMEs leverage AI platforms to integrate AI technologies? Ind. Mark. Manage. **107**, 466–483 (2022). https://doi.org/10.1016/j.indmarman.2022.10.008
50. Willenbacher, M., Scholten, J., Wohlgemuth, V.: Machine learning for optimization of energy and plastic consumption in the production of thermoplastic parts in SME. Sustainability (Switzerland) **13** (2021). https://doi.org/10.3390/su13126800
51. Willenbacher, M., Wohlgemuth, V., Risch, L.: Neural networks for energy optimization of production processes in small and medium sized enterprises. In: Wohlgemuth, V., Naumann, S., Behrens, G., Arndt, H.-K., Höb, M. (eds.) Advances and New Trends in Environmental Informatics: Environmental Informatics and the UN Sustainable Development Goals, pp. 129–145. Springer International Publishing, Cham (2023). https://doi.org/10.1007/978-3-031-18311-9_8
52. World Bank: Small and medium enterprises (SMEs) finance (2023). https://www.worldbank.org/en/topic/smefinance. Accessed 03 Feb 2025
53. Yang, X., Zhou, Z., Sørensen, J.H., Christensen, C.B., Ünalan, M., Zhang, X.: Automation of SME production with a Cobot system powered by learning-based vision. Robot. Comput.-Integr. Manufact. **83**, 102564 (2023). https://doi.org/10.1016/j.rcim.2023.102564

54. ZenBusiness: Zenbusiness takes the guesswork out of ai for small business owners (2024). https://www.businesswire.com/news/home/20241018984771/en/ZenBusiness-Takes-The-Guesswork-Out-Of-AI-For-Small-Business-Owners. Accessed 13 Jan 2025
55. Zhao, J., Zhang, L., Zhao, Y.: Informatization of accounting systems in small- and medium-sized enterprises based on artificial intelligence-enabled cloud computing. Comput. Intell. Neurosci. **2022** (2022). https://doi.org/10.1155/2022/6089195

Sentiment Classification of Product Influencers

Bárbara Freixo and Orlando Belo(✉)

ALGORITMI Research Centre / LASI, University of Minho, 4710-059 Braga, Portugal
pg49169@alunos.uminho.pt, obelo@di.uminho.pt

Abstract. Marketing Intelligence is a very attractive area for the application of sentiment analysis processes. Sentiment analysis can play an essential role in evaluating and optimizing influencer marketing in digital environments. Beyond the mere measurement of emotions, sentiment analysis can offer a deeper understanding of the impact of campaigns and influencers on audience perceptions. In this paper we present and discuss a sentiment analysis system conceived especially for identifying sentiments expressed in comments about influencers' posts placed on social networks. This can contribute to a deeper understanding of the level of customer satisfaction, their engagement, and perceptions of the content and products being promoted, among other things. The system was applied to a set of diversified comments about influencers' posts, which allowed for identifying the strengths and areas for improvement of each influencer.

Keywords: Decision Support Systems · Sentiment Analysis · Digital Influencers · Marketing Intelligence · Customer Profiling

1 Introduction

The evolution of social networks has significantly transformed the way companies promote their brands and products. With the exponential growth of these platforms, users use them as their primary source for information gathering, and especially for product researching. According to Beer [1], 54% of social media users do not actively publish, but use these platforms for searching information about people, products, or services. The global impact of social networks is undeniable. As of January 2020, 49% of the world's population was already using social networks [2], consolidating them as an indispensable channel for marketers. At the same time as the use of digital platforms has grown, the influence and relevance of digital influencers has also grown. These social media opinion agents can shape perceptions and influence purchasing decisions [3]. Influencers can attract loyalty followers, directly impacting their perceptions of brands and products [4–6]. This trend has boosted influencer marketing, an industry that has grown exponentially, reaching a value of $16.4 billion in 2022—an increase of more than 800% since 2016 [6].

Understanding the impact influencers have on brand perceptions has become increasingly essential for companies. According to a 2022 Nielsen report [8], nearly one in five Americans (19.3%) admit that a celebrity's endorsement influences their purchasing

B. Delibašić et al. (Eds.): ICDSST 2025, LNBIP 546, pp. 110–124, 2025.
https://doi.org/10.1007/978-3-031-90863-7_8

decisions. In addition, 71% of consumers trust product recommendations and opinions made by influencers, making them a key player in advertising campaigns. Influencers not only increase a brand's awareness but also boost purchase intent and affinity for it. It was also observed that 80% of consumers exposed to influencer-promoting ads remembered the brand presented, with 9% points increase in purchase intent among those who saw the ads versus those who were not. However, the influence is not limited to celebrities alone: 41.6% of consumers turn to the opinion of others before making purchase decisions, and 70% read online reviews before making a purchase. Given the immense number of comments generated daily on social media, it becomes essential for companies to implement automated systems for analyzing sentiments associated with the brand or the products they market or sell. This analysis offers very pertinent and valuable information. When properly explored, it can help in the definition of more effective marketing strategies, allowing companies to better understand their audience and adjust their campaigns based on the trends and perceptions identified, promoting communication more aligned with consumer expectations.

In this paper we present and discuss the development of a system capable of identifying the sentiments underlying comments on influencer posts placed on social networks. Analyzing these sentiments can contribute to a deeper understanding of the level of customer satisfaction, their engagement, and perceptions of the content and products being promoted. At an early stage, some existing models and techniques in the field of sentiment analysis were analyzed to define an approach that would ensure that a company's marketing strategies were aligned with these perceptions, following the trends and opinions of influencers' followers. Next, we define the application context to be explored and proceed to implement a specific process for collecting and preparing comments made in similar contexts by users of influencer publishing sites, which we previously selected. After normalizing the data collected, we selected a sample of influencers from the same market niche. The system was then applied to the comments on their posts, allowing them to identify the strengths and areas for improvement of each influencer. The remaining part of this paper is organized as follows. Section 2 approach several aspects of sentiment analysis and its application to real-world problems, Sect. 3, exposes the add-value of influencers for improving marketing strategies of companies, Sect. 4, presents the system developed, giving particular attention to its data preparation process, and Sect. 5 discusses sentiment classification and analyses the results obtained. Finally, Sect. 6 presents some conclusions and future work.

2 Analysing Sentiments

The way users use today the Web [9], and particularly social networks, differs greatly from what happened "half a dozen" years ago, not only due to the nature of services of the Web but also due to their computer literacy, regular Web experience and ease with computing platforms they have access to today. Currently, users not only consume Web content but also share their opinions and experiences about products they have purchased, services they have hired, or places they have visited, among many other things. These testimonies can significantly influence the decisions of other users, providing additional information and descriptions of very diverse contexts. The knowledge of these continuous, intensive, information exchange actions is valuable, both for companies, which can use this information to improve their products and services, and for the users themselves, who can benefit from the experiences of other users to make more thoughtful and sustained decisions. Everyone wins with this knowledge.

Sentiment analysis [10, 11] is a field of study specifically oriented towards the analysis of opinions, sentiments, evaluations, attitudes, and emotions of products, services, organizations, people, problems, events, topics, or characteristics. Sentiment analysis focuses primarily on opinions that express or imply feelings, whether these are positive, negative, or neutral. Other concepts, such as opinion analysis, opinion extraction, sentiment mining, subjectivity analysis and emotion analysis are covered by the domain of sentiment analysis and are often used as synonyms [12]. The origin of sentiment analysis is found in the disciplines of psychology, sociology, and anthropology, and comes from the theory of affective posture and the theory of evaluation, both focused on emotions as determinants in the formation of cognitions. Emotions are feelings generated from both conscious and unconscious processing. An emotional appraisal of a situation is a general appreciation of that situation, whether positive or negative, manifesting itself in mental and bodily responses [13]. In addition to positive and negative feelings, it is essential to consider expressions devoid of any implicit feelings, called neutral expressions. Sentences that express opinions are often characterized as subjective, unlike objective sentences that present facts. This is because opinions are intrinsically subjective. However, objective sentences can also present feelings, such as describing desirable or undesirable facts. In this way, sentiment analysis aims to identify feelings explicitly or implicitly expressed in texts, as well as the targets of these same feelings.

The "problem" of sentiment analysis can be understood as a vast set of interrelated sub-problems [14]. A single opinion is not enough to define sentiment about a specific entity or aspect, making it necessary to analyze a set of opinions. These can be expressed both informally and formally, including examples such as social media posts, product reviews, news, and forum discussions, which can vary in difficulty to analyze. Forum discussions, which are longer, more varied, and include interaction, typically represent a more complex approach compared to social media posts, which tend to be shorter and more direct. In addition, reviews can address a wide variety of topics, each presenting various levels of difficulty in their analysis. Predominantly, sentiment analysis identifies opinions that, in the form of subjective content, express positive or negative feelings, either explicitly or implicitly [12].

The effective implementation of a sentiment identification and classification system is a challenge with a high complexity. There are several approaches to its implementation, and there is no general process considered universal. A sentiment analysis system is not based on a single operation that performs the entire procedure. Instead, it is made up of a set of smaller tasks, each focused on analyzing specific pieces of information that we intend to analyze. However, Medhat et al. [15] proposed a general-purpose model for the implementation of sentiment analyses processes. This model integrates four basic tasks – identifying sentiments, selecting features, classifying sentiments, and defining sentiment polarity –, which the author referred as essential to any sentiment analysis process. All these tasks are chained and must be executed sequentially. Achieving high performance in each task individually is crucial to achieving high accuracy in the system.

3 The Add-Value of Influencers

The field of Marketing Intelligence [16] has been growing significantly in recent years, both in terms of analysis methods and of concrete applications. Several works have been developed by researchers, which recognized the usefulness of using sentiment analysis techniques for understanding consumer perceptions and reactions in marketing campaigns, especially those involving digital influencers. The use of influencers has become a frequent practice in marketing strategies, with brands taking advantage of their reach and engagement capacity to promote products and services. Sentiment analysis allows for monitoring, even in real time, the opinions of the followers of influencers, providing valuable information about the effectiveness of campaigns and the public's perception of brands or companies [17]. Sentiment analysis can play an essential role in evaluating and optimizing influencer marketing in digital environments, as it makes it possible to go beyond the mere measurement of emotions, offering a deeper understanding of the impact of campaigns and influencers on audience perceptions. By incorporating sentiment analysis into their marketing strategies, companies can make more informed decisions, mitigating risks and enhancing return on investment [18].

One of the biggest difficulties for companies is choosing the right influencer to promote their products or services. While an influencer may appear credible, their influence may be artificially inflated by bots or other fraudulent methods, which usually deceptively increase the number of followers. Sentiment analysis helps detect such situations by accurately alerting and monitoring the authenticity of interactions. An effective way to apply sentiment analysis in influencer marketing is by tracking specific topics relevant to a particular brand, from a particular company [19]. Over time, the application of sentiment analysis applied to specific topics can identify which influencers are generating positive responses and promoting discussions aligned with the company's values, allowing for a more strategic selection of partnerships. In addition to monitoring their topics of interest, businesses can also set up profile trackers for potential target influencers. This allows companies to choose influencers aligned with their goals, ensuring better promotional campaigns and avoid wasting resources.

A couple of years ago, Rambocas and Gama [13] presented a specific sentiment analysis process for the marketing area. The process proposed involves five distinct tasks, namely: data gathering, which consists of collecting opinions from various sources of information; pre-processing, which cleans and prepares the selected data for analysis; identification of feelings, in which the extraction of the feelings contained in the opinions is carried out; classification of feelings, which involves the categorization of feelings on a given scale of values; and, finally, presentation of results, in which the results of the sentiment analysis process are presented. Although, it is a very general process for sentiment analysis that was directed to the marketing area, because it includes the most common steps applied to the classification of polarity of feelings expressed in opinions, and thus on sentiment analysis classification processes.

4 Data Preparation

Often, the data we collect for analysis presents diverse types of consistency problems, such as null values, different formats, or heterogeneity of types. This usually compromises its utility. Therefore, it is essential to analyze and prepare them properly to ensure they are in good condition to be worked by analysis processes. The construction of a quality dataset, aligned with the objective of the analysis system, is especially important to ensure the success of the sentiment analysis process. A well-structured dataset, containing quality data, in adequate quantity and with the necessary diversity, is crucial to ensure the accuracy and robustness of the final process. For example, a poorly labeled dataset can lead to a machine learning model to identify incorrect patterns, resulting in inaccurate classifications for new data. In addition, if the dataset presents a large amount of data but is not truly diverse, the generalizability of the process may be limited. In this case, the process can become overly specialized in a single context, making it difficult to adapt to new situations. On the other hand, if the dataset presents a small amount of data, the process may struggle to learn meaningful patterns and will be more susceptible to overfitting issues.

Given the nature and goals of our application case, we focused our attention on social media, where product influencers are very active. In this domain, Instagram stands out as the main social network used by influencers, which has made it our main target for data gathering. However, after an exhaustive search for datasets suitable for the analysis process we wanted to carry out, we were unable to find data useful for us. This is because when we analyzed several influencers, having different numbers of followers, we found that, in all publications referring to product reviews, the influencer either expresses a positive opinion about the product or provides an objective description. This is because most of these publications are sponsored by companies. Thus, extracting only the information present in the influencer's description would not be useful or relevant, as it would result in a set of data with a lack of diversity and balance between classes, since there would be practically no negative feelings. So, we chose to build our dataset from scratch.

We started this process by extracting the comments from the users present in the influencers' posts, as these comments allowed us to access positive and neutral feelings, as well as a wide variety of negative feelings. This is something that we could not obtain from the sentiment expressed by the influencer alone. In addition, user comments tend to be more genuine, as they are not influenced by the company's sponsorship, as is the case with most opinions of the influencers who present it. To gather data, we developed a web scraper for extracting the comments of the users of influencers' publications. In the data extraction process, several publications from different influencers, who had different numbers of followers, were used, relating exclusively to the presentation of products. We only collect user feedback on public profiles of influencers. We did not collect any information about users, for respecting the General Data Protection Regulation. Then, we proceed to label the data we collected. This task was conducted manually, reading and labelling each data element – we decided to classify sentiments as positive, negative and neutral. To increase the diversity and quantity of the dataset, we created artificial data using GPT-4 [20]. Together with GPT-4, we used the prompt engineering approach [21], which provides model-specific textual instructions to generate new data in a controlled and targeted manner. The creation of the prompts was conducted with the objective of maintaining coherence, semantic relevance and variability in the dataset. At the end of the process, we obtained a set of 2484 comments, 55% real data and 45% fictitious data. All data extracted and created falls under the technology category. The comments were collected from publications about technological products made by influencers that are experts in technology. The final dataset is small. However, it is important to note that it was developed from scratch and labelled manually. As this is a research and study project, not a business project, we think that the size of the dataset is sufficient for the objectives of the work.

The dataset is organized into two attributes: "review", which represents the comments made by users, and "sentiment", which indicates the polarity corresponding to each comment. After a first analysis, we found that the dataset did not contain null values, which revealed that the developed Web scraper worked properly, and it was distributed in a very balanced way. This is a good characteristic, since sets of unbalanced data tend to favor the majority classes, compromising the accuracy of forecasts and generating biased analysis results, especially for the less represented classes. The distribution of the various elements of our dataset, in terms of polarity, was 35%, 33% and 32%, respectively for the neutral (0), positive (1) and negative (−1) class. Although the neutral class has a slight predominance, the difference is not significant enough to cause a relevant bias. In addition, the positive and negative classes are well balanced with each other, with a difference of only forty-six comments between them.

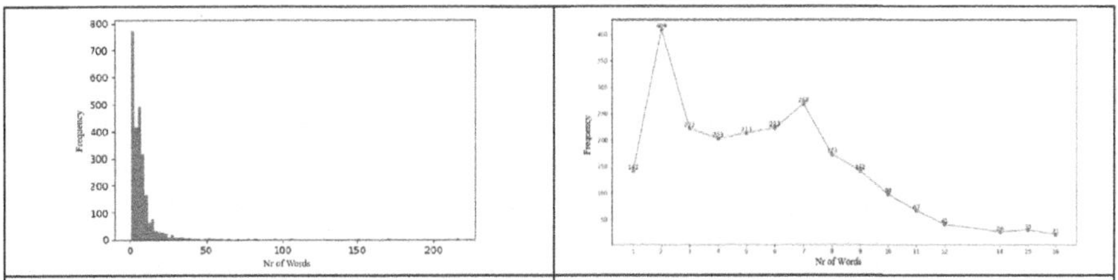

a) Distribution of words. | b) Number of occurrences per word.

Fig. 1. Distribution and occurrence of words in the dataset.

Next, we proceeded to analyze the number of words present in each comment, to obtain an initial insight into whether the removal of words with no associated sentimental value could contribute to increasing the accuracy of the models. With this analysis, we found that most of the comments are short, containing predominantly between 1 and 10 words (Fig. 1a). The distribution obtained is quite asymmetric, with a long "tail" on the right, which indicates that longer comments are relatively rare. Figure 1b shows us that the fifteen most frequent totals confirm this trend, showing a sharp drop in frequency as the number of words increases. These data show that short comments predominate, while longer ones are significantly less frequent. Removing words with no sentimental value may not always be beneficial, especially in short text contexts. Therefore, it is essential to use these techniques with caution, to avoid losing potentially relevant information.

To move forward in the analysis process, it was necessary to prepare the data according to the classification and sentiment analysis that we intended to perform. To do this, we have defined and implemented a specific data transformation process to standardize our initial dataset. The transformations of this process were carefully prepared, considering the nature of the comments: short length and noisy comments. In Fig. 2 we can see an illustration of the data transformation process. As we can see, this process can end in three different ways, producing three distinct datasets – DS1, DS2, and DS3. In practice, these three results derive from the application of the following transformation task flows: DS1) of not performing any transformation operations (flow 1), leaving the data as collected; DS2) to handle contractions and conversion of chat words (flow 2), followed by the usual tasks in a process of this kind; and DS3) after performing the process that produced DS2, perform the task of removing stop words (flow 3). Let us look at the various data transformation tasks embedded in that process.

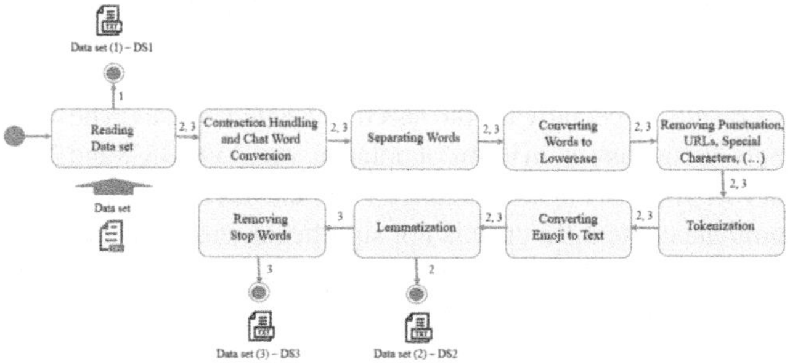

Fig. 2. The data transformation process.

One of the main characteristics of textual content on social networks is the frequent use of abbreviations and slang. It is important to treat these contractions, expanding abbreviated forms of words, transforming, for example, “don’t” into “do not” and “can’t” into “cannot”. This task contributes to vocabulary uniformity by reducing the number of different terms that models need to learn. Furthermore, many contractions can be interpreted in diverse ways depending on the context. For example, “he’d” can mean “he had” or “he would.” Expanding these contractions helps to clarify their meaning, thus reducing ambiguity in interpretations. Chat word conversion consists of transforming slang, abbreviations, and informal expressions, such as “u” (you) and “lol” (laughing out loud), into their full and formal forms. Converting these expressions also allows not only the uniformity of vocabulary but also contributes to the improvement of the quality of the data structure. Table 1 shows some examples of transformations that have been performed for treating contractions and converting chat words. This transformation had a small impact on the vocabulary size of our dataset. It eliminated about 5% of the initial records.

The quality of the syntactic structure of comments on social media posts is often compromised by problems such as disorganized writing, poor punctuation, and confusing syntax. As such, the quality of writing the dataset’s comments is one of the biggest challenges we face in analyzing it. Let us look at the following examples of comments present in the dataset:

Table 1. Contraction handling and chat word conversion example.

Review	Transformed review
I have the 4xl and I've been clutching it for years I love the features maybe I'll get the 7 pro now	*I have the 4xl and I have been clutching it for years I love the features maybe I will get the 7 pro now*
U like phone ♥	*you like phone* ♥

1) *So disappointing how hideous these phones look. The pixel 5 was absolutely beautiful. Completely ruined the design on the 6 and 7.*
2) *Last year Pixel's glass visor was too notch. They tried something new this time, which is decent, looks good, but in my opinion is a step back from a design point of view.*

As we can see from the underlined parts of the previous examples, there are several quality problems in their syntactic structures. Although humans can understand its true meaning, computers have difficulty doing so, treating these cases as if they were a single word. To solve these problems, a small script was developed that allowed the automatic identification of these cases, inserting the appropriate spaces and ensuring the correct separation of words. The data transformation process continues converting all the letters of the comments to lowercase for standardizing comments and ensuring that the same words, spelled differently, such as "Good" and "good", are treated as false. Without this transformation, equal words would be treated as distinct. This has a significant impact on solutions based on supervised learning, reducing the vocabulary of the dataset and the dimensionality of the analysis vectors, as well as eliminating any redundancies. With this task the dataset was reduced by about 15% more. Next, punctuation, numbers, mentions, URLs, and special characters are eliminated. The identification of sentiments in comments is determined by the words that contain an associated sentimental value. These are fundamental, while other elements of the text can add noise, impairing clarity, data quality, and the consequent accuracy of the results. It is essential to highlight the words of feeling. This involves the removal of unnecessary terms, which do not contribute to the determination of polarity.

Tokenization is the next task. When a comment is tokenized, the text is segmented into tokens, which correspond to individual words. This process involves separating the text, resulting in a list of words or terms that represent the original text. This transformation makes it easier to analyze each word individually in vocabulary. After tokenization, we converted emojis to text. The increasing use of emojis on social networks reflects their growing importance in expressing feelings. They have become widely adopted to communicate emotions, reactions, and nuances that are often difficult to express using words. Therefore, it is important to treat them with particular care. While some models are capable of handling emojis in their original form, many others are not optimized to interpret these symbols directly. Emojis conversion allows sentiment analysis models to use their textual analysis capacity to process and understand their content. Usually, the vocabulary of languages has words having several variations. For example, nouns can vary in plurality, while verbs can vary according to the tense used. In many of these cases, variations do not add additional knowledge to the learning models. To eliminate variations, we use lemmatization for transforming them into their base form, or lemma. Removing stop words is the next task. Stop words are considered irrelevant to the text because, although they are frequently used, they do not carry substantial meaning. However, the decision to remove them may vary depending on the analysis process. Removing these words should be evaluated based on the impact they may have on the data, considering whether they are important for processing and obtaining accurate results. This is the last data transformation task we implemented.

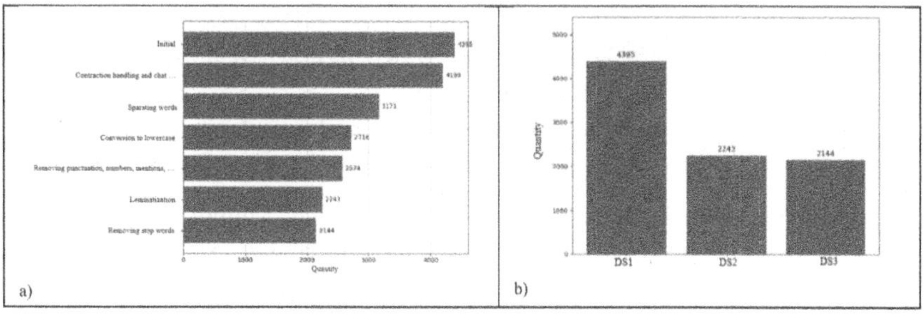

Fig. 3. The evolution of the dataset.

The data transformation process allowed us to normalize and reduce the complexity of the initial dataset. This was achieved through transformations that decreased the size of the dataset's vocabulary. However, it is critical to understand the nature of the data itself. The fact that the dataset is mostly composed of short comments required a cautious reduction of vocabulary. In the next section we will see which transformations were the most appropriate to achieve the objectives of this work. It is important to note that each system must use the transformations that best suit its characteristics, since there is no universal set of steps applicable to all cases. In Fig. 3a we can see how the dataset was reduced after each transformation and in Fig. 3b the size of the various datasets generated by each flow (1, 2 and 3) of the data transformation process.

5 Classifying Sentiments

After finishing the data preparation process and normalizing the datasets (DS1, DS2 and DS3), we moved on to implementing the sentiment classification process of the comments. To carry out this process, we can use diverse types of models, whether based on machine learning, lexicons, or hybrid approaches. However, supervised machine learning models are more suitable for working data from social networks, as they can identify complex patterns and capture linguistic nuances, such as sarcasm or slang, and can adapt quickly to new language trends. Thus, we decided to develop a set of supervised machine learning models to do the desired sentiment classification.

5.1 Supervised Learning Approaches

The development of a supervised learning solution requires testing several classifiers, for identifying the one that best fits the final analysis system based on the quality of the results, according to the metrics defined for the evaluation of the models. In addition to the use of various classifiers, we also can implement different techniques to improve further the quality of the models. Among these techniques are dataset balancing methods, such as oversampling and undersampling, as well as attribute extraction techniques. Despite the complexity of this process, it is essential to ensure the selection of the most efficient model. The techniques and models we selected as well as the classifiers we used in this work are presented in Table 2.

Table 2. Techniques, models and classifiers used.

Techniques and Models	Classifiers
Oversampling + TF-IDF Vectorizer	Naive Bayes, Random Forest, XGBoost, Support Vector Machines, and Logistic Regression
Oversampling + Bag of Words	Naive Bayes, Random Forest, XGBoost, Support Vector Machines and Logistic Regression

To compare the performance of the models, it is essential to select an appropriate evaluation metric. Knowing that our goal is to predict the sentiments expressed in comments about influencers' posts, it is critical to get the highest possible ratio of correct predictions out of the total predictions made. To do this, we used accuracy as a metric. Accuracy is defined as the ratio of the number of correctly rated reviews to the total ratings made, which allows for measuring the accuracy of predictions by correctly identifying feedback sentiment. It is also important to divide the data between training and testing. For this, we used the train-test split technique, which performs a random division of the data. In this case, 75% of the comments were intended for training and 25% for testing. All models developed used the same training and testing set, ensuring consistency in the evaluation process. Furthermore, the training set was balanced between the different classes to ensure that the models had an adequate representation of the examples of each class. In this way, models can learn in a balanced way, avoiding focusing excessively on the classes with the most examples, while ensuring the correct learning of the minority classes. Although the dataset classes are relatively evenly distributed, it is beneficial to balance the dataset, because of the very limited number of comments available it has. Thus, to increase the number of comments in minority classes, we used oversampling to generate new data for classes with lower representation. We used SMOTE (Synthetic Minority Over-sampling Technique) [22] to do this. Unlike direct duplication of data elements, this method allows us to create new synthetic comments, generating interpolated data from the existing ones. In this way, we were able to avoid possible overfitting problems, which could occur if we opted for direct duplication of comments. In our case, the balance was only applied to the training set. Doing this we ensured that the model is trained with balanced data, while the test set remains unchanged, preserving the original characteristics of comments. Thus, the models are guaranteed to be evaluated with original data without the occurrence of artificial data resulting from the application of SMOTE. Before applying balancing, we remove duplicate data from the training set, minimizing the risk of overfitting. After removing the duplicate data and balancing the training set, it is critical to turn the comments into representative vectors to make supervised models easier to learn. To do this, we used extraction techniques, such as TF-IDF Vectorizer and Bag of Words, which convert each comment into numerical vectors, which allows models to process them efficiently and better understand the patterns present in data. The vectors generated by the attribute extraction techniques were used to train and evaluate the different supervised models. Finally, to obtain the parameters of the classifiers automatically, we used GridSearchCV, which ensured the best selection of parameters for each model developed. A GridSearchCV evaluates all possible combinations of a pre-defined set of parameters for each classifier used, selecting the most suitable classifiers according to the chosen metric. Table 3 shows the accuracy values we got for each model implemented.

Table 3. The accuracy of the models.

Model	DS1	DS2	DS3
Oversampling + TF-IDF Vectorizer + Support Vector Machine	0.71	0.83	0.8
Oversampling + TF-IDF Vectorizer + Random Forest	0.68	0.78	0.75
Oversampling + TF-IDF Vectorizer + Naïve Bayes	0.68	0.79	0.77
Oversampling + TF-IDF Vectorizer + Logistic Regression	0.72	0.83	0.79
Oversampling + TF-IDF Vectorizer + XGBoost	0.64	0.77	0.74
Oversampling + Bag of Words + Support Vector Machine	0.67	0.81	0.81
Oversampling + Bag of Words + Random Forest	0.68	0.75	0.75
Oversampling + Bag of Words + Naïve Bayes	0.68	0.78	0.77
Oversampling + Bag of Words + Logistic Regression	0.70	0.81	0.81
Oversampling + Bag of Words + XGBoost	0.66	0.8	0.79

As we can see, the DS1 data set was the one that obtained the worst results, while the results obtained for the DS2 data set are slightly higher than those of the DS3 data set, with an increase of about 17%, both for the cases using TF-IDF Vectorizer for attribute extraction and for those using Bag of Words. When we compare the results of the DS3 data set, we see an average improvement of about 12% with TF-IDF Vectorizer and approximately 16% with Bag of Words. In both cases, using TF-IDF Vectorizer and Bag of Words, the DS2 data set provided better results. However, the difference between the results of the DS2 and DS3 data set is more pronounced when using TF-IDF Vectorizer, compared to Bag of Words. With TF-IDF Vectorizer, the difference varies between 2 and 4% points, while with Bag of Words the results vary between 0 and 1% points. This suggests that not removing stop words significantly benefits the accuracy of models with TF-IDF Vectorizer, while in the case of Bag of Words the improvement is more subtle. An explanation for this observation is the fact that TF-IDF Vectorizer assigns greater weight to less frequent words in the corpus, minimizing the impact of stop words when they are not removed, which allows a better distinction between informational terms. In contrast, the bag of words is limited to counting the frequency of words, without considering their relevance, which makes stop words dilute the importance of the most significant terms. Thus, in our case, removing stop words is not beneficial. This can be justified by the fact that stop words provide essential context in short and informal comments, help to capture the syntactic relationship between words and preserve colloquial expressions, which are fundamental for the correct interpretation of sentiments. Moreover, sentiment modifiers can invert or reinforce sentiments, making their removal detrimental to the analysis. In Fig. 4, we can see the confusion matrix obtained for our best model, considering test data.

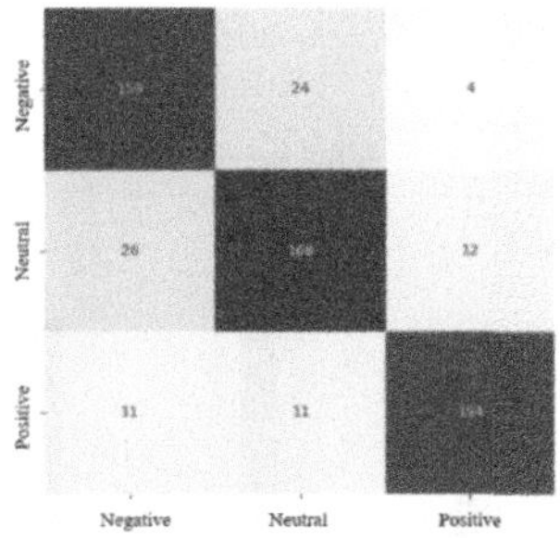

Fig. 4. The confusion matrix of the best model.

Observing the confusion matrix, we can see that the greatest incorrect predictions occur between the neutral and negative classes. About 57% of classification errors involve these two classes. This may be because neutral reviews often have greater ambiguity, making them more difficult to classify accurately. In addition, negative comments can contain elements of sarcasm, which often make it difficult for humans and classification models to interpret them. To improve the model, it is essential to focus on improving the quality and increasing the quantity of data we used, ensuring greater diversity. This will help reduce the average confidence in incorrect predictions and consequently improve the accuracy of the model.

5.2 Classifying Sentiments of Product Influencers

Once the sentiment analysis model was defined, we moved on to the analysis phase of influencer publications, with the aim of studying user interactions, identifying patterns of behavior, and determining which influencers have the greatest impact on the promotion of certain products. This process allows us to identify the most relevant influencers and understand the influence they have on their followers. This can effectively contribute to ensuring more thoughtful decision-making in marketing campaigns. The classification model we used to make this evaluation was the one that presented the best accuracy in our evaluation process: the "TF-IDF Vectorizer + SentiWordNet + Bigrams + Support Vector Machine", using the data preparation model of the DS2 data set. For doing this analysis, we decided to evaluate three influencers (A, B and C) from the technology product segment. The process was executed in a similar way to the one we used in the preparation of the data set DS2 (Sect. 4), performing all the preparation tasks in the same way, but focusing the process and data gathering on the social network areas of these influencers. Table 4 presents a summary of the data collected, for each influencer. After processing, influencers' data was classified with our best model. The results of the classification process are presented in Fig. 5.

Table 4. Influencers data characteristics.

Influencer	Followers	Average of Comments	Likes
A	1400000	94	3634
B	494000	40	4911
C	288000	67	7799

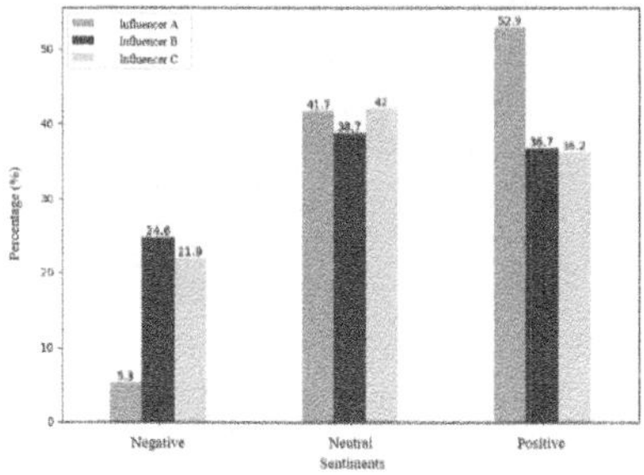

Fig. 5. Sentiment classification results.

From the classification data we can see that influencer A stands out for the predominance of positive sentiments – 52.9% of feelings are positive, and only 5.3% are negative. The high proportion of neutral comments (41.7%) suggest that several sentiments are not related to the content, indicating that there is considerable interaction, but the comments made are not relevant. Influencer B has much less satisfactory results, presenting a considerable proportion of negative feelings (24.6%) and only 36.7% of positive ones, which indicates a more divided perception. The high proportion of neutral feelings (38.7%) suggests that many comments are also superficial or out of context. Finally, influencer C has a distinct dynamic, with a similar proportion of positive sentiments (36.2%) as influencer B, but with a slightly lower rate of negative sentiments (21.9%). The high proportion of neutrals (42.0%) suggests that a lot of comments are also off-topic, as is the case with Influencers A and B. Our analysis process went a step further, covering more specific aspects. However, given the current size of this article, it is not possible to include these results here.

6 Conclusions and Future Work

In this paper, we presented and discussed a sentiment classification and analysis process we carried on over data containing comments about posts that product influencers published on their social networks accounts. Our goal was to get a deeper understanding of the level of customer satisfaction, their engagement, and perceptions of the content and products being promoted from some product influencers. The system was applied to a set of diversified comments about influencers' posts, which allowed get important knowledge for identifying the strengths and areas for improvement of each influencer.

In the sentiment classification process, we implemented various supervised machine learning models. As we know, each problem has its specificities. There are no universal solutions. With that in mind, we implemented and assessed different classifiers to determine which classification model would be best suited for our purposes. Since no classifier can process textual content in its original format, we used various attribute extraction techniques, such as TF-IDF Vectorizer and Bag of Words to convert the texts into representative vectors. In total, we used five classifiers along with these two techniques and with a data balancing technique, SMOTE. The first classifications showed very satisfactory results, but we sought to improve the performance even more. To do this, we use bigrams and the SentiWordNet lexicon, adding an additional semantic layer to the attribute extraction techniques. In this way, we improve the results of the models we implemented. We believe the results we got are quite satisfying, especially considering the limitations encountered throughout the process, particularly in data gathering operations in terms of quality and quantity of comments. Even not compromising the goals established previously or the implementation of the classification system, such limitations offer clear opportunities to improve some aspects of the work carried out, in particular: expanding the dataset used, including a greater number of publications from influencers from different niches, ensuring a more diverse and representative dataset for analysis, which may contribute to greater precision and generalization of the results as well as to make the system more robust; improve data labelling quality by ensuring that sentiments are correctly classified – more accurate labelling has an impact on the effectiveness of machine learning models; and, finally, automate the process of selecting influencers from a certain niche, specializing the sentiment analysis process that we have now. Furthermore, we intend to evaluate the use of more sophisticated natural language processing techniques and methods, as well as upgrade the system we developed to receive advanced deep learning techniques for refining sentiment analysis and classification.

Acknowledgements. This work has been supported by FCT – Fundação para a Ciência e Tecnologia within the R&D Unit Project Scope UID/00319/Centro ALGORITMI (ALGORITMI/UM).

References

1. Beer, C.: Social Browsers Engage with Brands. GWI (2018). https://www.gwi.com/blog/social-browsers-brand
2. Statista: Number of social network users worldwide from 2010 to 2023 (2020). https://www.statista.com/statistics/278414/number-of-worldwide-social-network-users/
3. Watts, D.J., Dodds, P.S.: Influentials, networks, and public opinion formation. J. Consum. Res. **34**(4), 441–458 (2007)
4. De Veirman, M., Cauberghe, V., Hudders, L.: Marketing through Instagram influencers: the impact of number of followers and product divergence on brand attitude. Int. J. Advert. **36**(5), 798–828 (2017)
5. Colliander, J., Dahlén, M.: Following the fashionable friend: the power of social media—Weighing the publicity effectiveness of blogs versus online magazines. J. Advertising Res. **51**(1), 313–320 (2011)

6. Lee, J.E., Watkins, B.: YouTube vloggers' influence on consumer luxury brand perceptions and intentions. J. Bus. Res. **69**(12), 5753–5760 (2016)
7. Santora, J.: Key influencer marketing statistics you need to know for 2022. Influencer Marketing Hub (2022). https://influencermarketinghub.com/influencer-marketing-statistics/
8. Nielsen: How marketers can successfully leverage social media influencers in their campaigns. Nielsen (2022). https://www.nielsen.com/insights/2022/how-marketers-can-successfully-leverage-social-media-influencers-in-their-campaigns/
9. Palomino, F., Paz, F., Moquillaza, A.: Web analytics for user experience: a systematic literature review. In: Soares, M.M., Rosenzweig, E., Marcus, A. (eds.) Design, User Experience, and Usability: UX Research and Design. HCII 2021. LNCS, vol. 12779, pp. 312–326. Springer, Cham (2021). https://doi.org/10.1007/978-3-030-78221-4_21
10. Mohey El-Din, D.: A survey on sentiment analysis challenges. J. King Saud Univ. Eng. Sci. (2016). https://doi.org/10.1016/j.jksues.2016.04.002
11. Routray, P., Swain, C., Mishra, S.: A survey on sentiment analysis. Int. J. Comput. Appl. **76**(10) (2013). (0975–8887)
12. Liu, B.: Sentiment analysis and opinion mining. Synth. Lect. Hum. Lang. Technol. **5**(1), 1–184 (2012). https://doi.org/10.1007/978-3-031-02145-9
13. Rambocas, M., Gama, J.: Marketing Research: The Role of Sentiment Analysis. Working Papers (FEP) - Universidade Do Porto (2013)
14. Verma, B., Thakur, R.S.: Sentiment analysis using lexicon and machine learning-based approaches: a survey. In: Tiwari, B., Tiwari, V., Das, K., Mishra, D., Bansal, J. (eds.) Proceedings of International Conference on Recent Advancement on Computer and Communication. LNNS, vol. 34, pp. 441–447. Springer, Singapore (2018). https://doi.org/10.1007/978-981-10-8198-9_46
15. Medhat, W., Hassan, A., Korashy, H.: Sentiment analysis algorithms and applications: a survey. Ain Shams Eng. J. **5**(4), 1093–1113 (2014)
16. Mandal, P.C.: Marketing information and marketing intelligence: generation of customer insights. Int. J. Technol. Diffusion (IJTD) **13**(1), 1–14 (2022). https://doi.org/10.4018/IJTD.300745
17. Hughes, C., Swaminathan, V., Brooks, G.: Driving brand engagement through online social influencers: an empirical investigation of sponsored blogging campaigns. J. Mark. **83**(5), 78–96 (2019)
18. Agarwal, S., Damle, M.: Sentiment analysis to evaluate influencer marketing: exploring to identify the parameters of influence. PalArch's J. Archaeol. Egypt/Egyptol. **17**(6), 4784–4800 (2020)
19. Rambocas, M., Pacheco, B.G.: Online sentiment analysis in marketing research: a review. J. Res. Interact. Mark. **12**(2), 146–163 (2018)
20. OpenAI et al.: GPT-4 Technical report (2024). https://doi.org/10.48550/arXiv.2303.08774
21. Ekin, S.: Prompt Engineering For ChatGPT: A Quick Guide To Techniques, Tips, And Best Practices. TechRxiv, 04 May (2023). https://doi.org/10.36227/techrxiv.22683919.v2
22. Chawla, V., Bowyer, W., Hall, O., Kegelmeyer, P.: SMOTE: synthetic minority over-sampling technique. J. Artif. Intell. Res. **16**, 321–357 (2002). https://doi.org/10.1613/jair.953

Decision Support System Challenges

A System Dynamics Model to Promote Awareness of the Green Transition Challenges in Higher Education Institutions

Eduardo Franco[1](✉), Stefano Armenia[2], Odysseas Christou[3], George Tsaples[4], Juan Uribe-Toril[5], and Jason Papathanasiou[4]

[1] University of São Paulo, São Paulo, Brazil
eduardo.franco@alumni.usp.br
[2] Università Telematica degli Studi, Florence, Italy
s.armenia@iuline.it
[3] University of Nicosia, Nicosia, Cyprus
christou.o@unic.ac.cy
[4] University of Macedonia, Thessaloniki, Greece
{gtsaples,jasonp}@uom.edu.gr
[5] University of Almeria, Almeria, Spain
juribe@ual.es

Abstract. Higher education institutions (HEIs) are crucial in transitioning to sustainable, carbon-neutral societies. However, achieving sustainability in HEIs requires systemic transformation across governance, operations, teaching, and research. This paper presents a simulation model designed to enhance awareness and strategic decision-making regarding sustainability in HEIs. The model was developed within a European transnational project focusing on a Whole Institution Approach (WIA) to sustainability through systems thinking. By integrating feedback loops and dynamic interdependencies, the model allows HEIs to simulate and evaluate policies, assess long-term sustainability impacts, and identify leverage points for transformation. A hypothetical case study illustrates the model's application, providing an interactive learning environment for stakeholders. The model highlights the importance of aligning policies, resource allocation, and cultural shifts within HEIs to address the green transition and stimulates reflection on mental models and the development of sustainability strategies.

Keywords: System Dynamics · Higher Education · Sustainability · Whole Institution Approach · Decision Support Model

1 Introduction

Environmental challenges are economic in nature but also have societal implications. From a system perspective, sustainable development cannot be achieved by only examining the environmental dimension without considering the economic and social dimensions [1]. Higher education institutions (HEIs) are essential to transitioning towards

B. Delibašić et al. (Eds.): ICDSST 2025, LNBIP 546, pp. 127–138, 2025.
https://doi.org/10.1007/978-3-031-90863-7_9

carbon neutrality and sustainable societies [2]. Besides, HEI is a place capable of stimulating all its actors to reflect, discuss, and understand what is wrong with the whole system and identify solutions and alternatives to fix it. Still, education stakeholders cannot do that in isolation from the community and society in which they live.

HEIs can be an experiential place of learning for sustainability and should, therefore, orient all their processes toward sustainability principles. Educational institutions must be transformed, and mainstream sustainability should be integrated into all aspects [3]. In practice, a whole-institution approach suggests incorporating sustainable development through curriculum elements and integrated management and governance, making the institutions into microcosms of sustainability [4].

Data collected by a survey of greening in higher education in Europe, conducted by the European University Association (EUA) [5], provided evidence that environmental sustainability is already of high importance at the central level of HEIs, with three-quarters of those institutions surveyed across Europe have taken concrete steps to address environmental sustainability at the strategic level. However, drivers and approaches can be different across institutions. Most HEIs lack dedicated policies and strategies that comprehensively and holistically address all aspects of greening and sustainability.

UNESCO [6] suggested areas in which the whole-institution approach could be realized within the HEI context, including:

a) An institution-wide process enables all stakeholders—leadership, teachers, learners, and administration—to jointly develop a vision and plan to implement sustainability throughout the institution.
b) Technical and, where possible and appropriate, financial support is provided to the institution to support its reorientation. This can include providing relevant good practice examples, leadership and administration training, guidelines development, and associated research.
c) Existing relevant inter-institutional networks are mobilized and enhanced to facilitate mutual support, such as peer-to-peer learning on a whole-institution approach, and to increase the visibility of the approach to promote it as a model for adaptation.

The evolution of HEIs into more sustainable institutions means they are more resilient and better prepared for new challenges. The proposed model also has the potential to be adapted to other education sectors (e.g., primary, secondary, VET, and adult).

As problems and challenges become more complex and interlinked, an integrated and systemic approach is necessary. For the socio-technical complex challenge HEIs face while pursuing the green transition, an approach combining systems thinking, knowledge from domain experts, and eliciting mental models for developing a tool to support consensus building and the decision-making process could help [7].

The present study presents an interactive learning environment (ILE), based on a system dynamics simulation model, to support HEIs in building stakeholder awareness. The model could also support strategic planning on effectively responding to evolving needs and conditions to deliver systemic, institution-wide integration of sustainability.

The proposed ILE and simulation model was developed in the context of a European transnational project entitled "Fostering whole institution approach to sustainability in

HE through systems thinking,"[1] which seeks to identify, develop, test, and assess the whole institution's approach to sustainability through systems thinking in the HE sector. Thus, it responds to the vital European priorities of the European Green Deal [8, 9] and calls for education and training systems and institutions to become catalysts for the transition to a greener and more sustainable Europe.

2 Background

2.1 Systems Approaches

A "system" can be defined as a collection of interconnected and interdependent components or elements that work together to achieve a common purpose or goal [10]. These components can be tangible entities such as physical objects, people, or organizations and intangible elements such as processes, information, or ideas.

Schwaninger [11] described "the system approach" as a framework based on system theory (the science of the structure and behavior of organized wholes) and cybernetics (the science of control and communication in complex dynamical systems). This framework provides a formal apparatus for dealing with complex systems from many fields and allows synergic interaction between disciplines.

Systems thinking is a holistic approach to understanding and analyzing complex systems. Rather than focusing on individual parts in isolation, systems thinking emphasizes the interconnectedness and interdependence among these parts. It recognizes that changes in one part of a system can have far-reaching effects on other parts and the system as a whole [12]. It seeks to identify underlying patterns, structures, and feedback loops that influence behaviors and outcomes [13].

System dynamics is an approach to studying and understanding complex systems that emphasizes the interrelationships and dynamic behavior of the system's components. Forrester developed it in the 1960s [14–16] and has been widely applied in various fields, including management, economics, and social sciences.

System dynamics uses computer simulation models to represent and analyze systems' behavior over time and focuses on understanding how the interactions between components, feedback loops, and time delays contribute to the system's overall complex behavior and outcomes. It is founded on the scientific method, and it consists of an iterative approach that seeks to 1) articulate a problem, 2) formulate a dynamic hypothesis, 3) develop a simulation model, 4) test and build confidence in the proposed formulation, and then 5) design and evaluate intervention policies. Sterman [17] provided a good description of its applicability:

> "[It] enables us to build formal computer simulations of complex systems and use them to design more effective policies and organizations. Together, these tools allow us to create management flight simulators-microworlds where space and time can be compressed and slowed so we can experience the long-term side effects of decisions, speed learning, develop our understanding of complex systems, and design structures and strategies for greater success."

[1] https://susthein.com/.

One key activity in this approach is modeling feedback mechanisms during the simulation model development. Most complex behaviors usually arise from the interaction of two basic feedback loop types: balancing and reinforcing. Capturing those intricacies contributes to a better understanding of the problem under investigation, and by combining feedback loops, it is possible to create causal loop diagrams (Fig. 1).

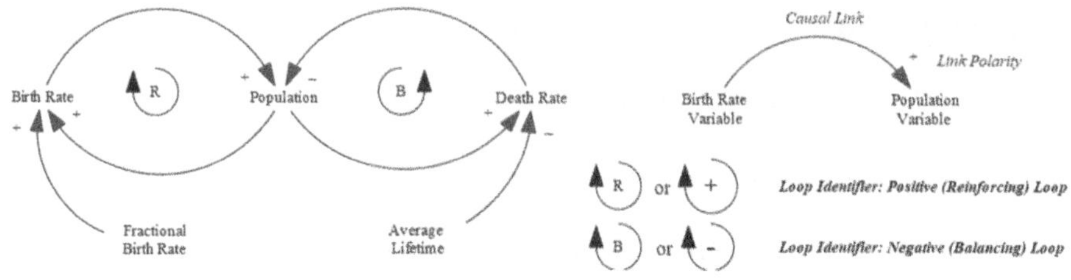

Fig. 1. Examples of the elements of a causal loop diagram, adapted from [17]

Following the qualitative modeling through the causal loop diagram, the implementation of a system dynamics simulation model is based on a set of coupled, nonlinear, and first-order differential equations, which can be represented as follows [17]:

$$\dot{x}(t) = f(x, p)$$

where "x" is a vector of state variables, "p" is a set of parameters, and "f" is a nonlinear vector-valued function.

2.2 Whole Institution Approach and Sustainable Development in HEI

The whole-system approach provides a framework for addressing complex problems by considering the interconnectedness and interdependencies of various components within a system. It promotes a comprehensive and integrated understanding of complex challenges and helps design effective and sustainable policies and strategies.

Blizzard and Klotz [18] proposed the following definition:

> "Whole systems design considers an entire system as a whole from multiple perspectives to understand how its parts can work together as a system to create synergies and solve multiple design problems simultaneously. It is an interdisciplinary, collaborative, and iterative process."

A significant aspect of the WIA is its focus on sustainability. Menon and Suresh [19] highlighted how sustainability practices are integrated into education, research, campus operations, and community engagement. This integration aligns with the UN's decade of education for sustainable development, which has driven institutions to adopt holistic approaches encompassing various operations. Similarly, Weiss and Barth [20] conducted a systematic review that reveals the implementation processes of sustainability curricula in higher education, emphasizing the need for a comprehensive approach that acknowledges sustainability as a cross-cutting theme in educational practices.

The WIA emphasizes the need for a comprehensive integration of sustainability across all institutional functions, including curriculum, research, operations, and community engagement. Hargis et al. [21] argued that WIA contributes to climate change education, encouraging collaborative action rather than isolated responses. This perspective is supported by Kohl et al. [22], who highlighted that HEIs can contribute to the Sustainable Development Goals[2] (SDGs) by embedding sustainability into their core operations through a whole-institution approach and, thus, transforming institutions into catalysts for societal change, aligning educational practices with sustainability objectives.

WIA focuses on systemic change, which is crucial in addressing the rapidly evolving demands of higher education. Chow [23] highlighted that systemic change provides a holistic perspective for assessing organizational needs at multiple levels, thus enabling institutions to adapt to changing educational landscapes.

The whole institution approach in higher education emphasizes the integration of various components within educational institutions to enhance academic quality, sustainability, and institutional effectiveness. This approach recognizes that the effectiveness of HEIs is not solely dependent on individual programs or departments but rather on the interconnectedness of all elements within the institution, including governance, curriculum, faculty members' development, and student services.

3 Development of the Simulation Model

Combining systems approaches, especially systems thinking, system dynamics, and group model building (GMB), can be highly beneficial in creating practical guidelines for sustainability policies in higher education institutions (HEIs) [24, 25].

Systems thinking can encourage stakeholders to view HEIs as complex, interdependent systems where changes in one area can have ripple effects throughout the institution. This perspective is crucial for understanding how different components of an institution—such as academics, students, operations, and administration—interact and influence each other. By adopting systems thinking, stakeholders can identify leverage points within the institution that can drive significant improvements in sustainability. This holistic approach ensures that sustainability efforts are not isolated but integrated across all institution levels.

System dynamics involves creating simulation models representing the institutional system's behavior over time. This allows stakeholders to test different scenarios and predict the outcomes of various policy decisions.

Group Model Building (GMB) is a participatory process that engages stakeholders in developing causal representations and system dynamics models. This approach is efficient in complex settings involving multiple stakeholders, as it facilitates a collaborative understanding of the system in question and the interconnections between its components [26, 27]. Involving stakeholders in the modeling process helps to ensure that diverse perspectives are considered, enhancing the relevance and accuracy of the model [28].

[2] https://sdgs.un.org/.

The participating HEIs in the research project developed a comprehensive simulation model by integrating these methodologies. Two GMB sessions were organized during the initial steps of the modeling phase (Fig. 2). Through facilitated workshops, stakeholders (internal HEI staff, members from the project team, and external domain experts) contributed their knowledge and perspectives, ensuring the model reflected the institution's real-world complexities.

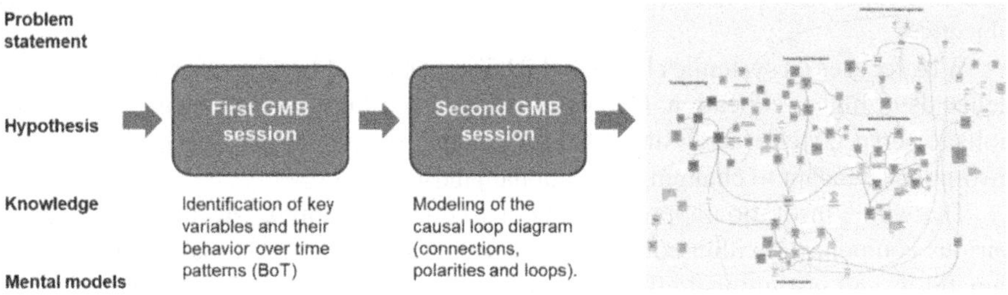

Fig. 2. Group-Model Building sessions' inputs and outputs.

The two GMB sessions produced an initial version of the causal loop diagram (CLD), which underwent further analysis and revisions during consultation meetings with invited external experts and project partners. Figure 3 shows the overview of the GMB output, considering the key subsystems identified and their interrelationships.

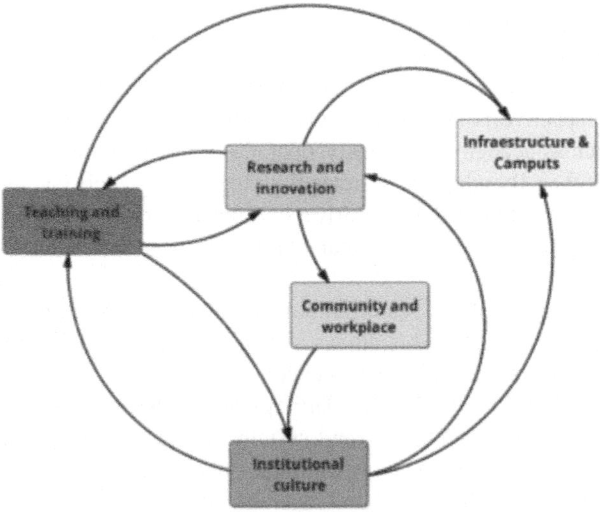

Fig. 3. The subsystem diagram provides an overview of the developed CLD.

Following the GMB session, the simulation model was developed, and its formulation was submitted for analysis and feedback from subject matter experts outside the research project team. Those inputs provided valuable insights that were used to improve the initial formulation and the developed interactive simulation model.

The Stella Architect [29] was used to develop the simulation model and its web user interface, which was then uploaded to an online simulation model library[3].

4 The Interactive Simulation Model

The developed simulation model considers a hypothetical scenario for a role-playing game where users can test their decisions and learn from the outcomes obtained from their choices. This hypothetical scenario is based on secondary data from several sources, and it is called the "Green Future University" (GFU), consisting of a small higher education institution with 1,000 students, 1 staff member per 10 students, and 1 faculty member per 7 students, situated on a compact 10,000-square-meter campus. Despite its potential, GFU faces significant sustainability challenges, with outdated, unsustainable infrastructure and limited resources to address the green transition challenges.

This case study invites participants to reimagine GFU as a model of green transformation through simulation and role-playing. The users are welcome to tackle real-world challenges, such as retrofitting infrastructure, curricula transformation, campus operation, aligning stakeholders' interests, and fostering a culture of sustainability, all while navigating the constraints of a small yet ambitious institution.

The user's goal is to play the role of the rector and lead GFU's transition into a sustainable future. As rector, the user is responsible for making critical decisions on infrastructure development (choose between green or regular upgrades to meet campus needs), faculty and staff ratios (balance the number of educators and staff to ensure student success), sustainability culture (promote sustainable behaviors to foster a green campus mindset); and a very tight budget allocation (strategically distribute financial resources across various operational and developmental priorities).

The model formulation and KPIs were organized following the structures and causal relationships identified in the GMB sessions and then refined during the model development process. Table 1 summarizes the main concepts, structures, and dynamics of the four subsystems implemented in the model. The fifth dimension, i.e., the institutional culture, is represented by the mental model of the users playing with the simulation model and their decisions.

After presenting the user with the overall context and objective of the role-playing simulation, the user navigates to the simulation dashboard, where they can control the simulation by making decisions on managing the hypothetical higher education institution. Figure 4 shows one of the simulation model user interfaces, consisting of the dashboard where the user analyzes the GFU's current performance through a set of KPIs about subsystems previously presented in Table 1.

[3] https://exchange.iseesystems.com/public/efranco/sust-hein---the-green-future-university-case/index.html.

Table 1. Description of the simulation model subsystems

Subsystem	Description
Community and workplace	There are three types of populations: students, faculty members, and staff members. All can increase and decrease over time as conditions are appropriate or inappropriate. Each is divided into two subsets: those aware of sustainability issues and those unaware, which influence their behavior. They follow the innovation diffusion concept: based on some conditions, such as classes, dissemination campaigns, and word-of-mouth, they become aware of or return to the initial state of unawareness.
Campus infrastructure	On average, 10 m^2 is necessary for each enrolled student in the university (considering all types of buildings and facilities). That infrastructure can be either sustainable or not, and sustainable infrastructure is 15% more expensive to build than traditional. The infrastructure directly influences the operational environmental footprints: energy consumption, waste generation, water consumption, and greenhouse gas emissions. Building new infrastructures takes at least three years, and renovating old and traditional infrastructures to sustainable infrastructures is possible.
Teaching and training	The faculty member-per-student ratio determines the student capacity of each course. In addition to regular classes, other short training activities are created within research projects and are offered to the broad HEI community. Both classes and training activities can be sustainability-oriented or not, depending on the profiles of the faculty members and the ongoing research projects. They can change the behavior of the university population toward sustainability goals as they are exposed to them, and if not, return to their non-sustainable behaviors.
Research and innovation	Research proposals are created proportionally to the number of faculty members and determined by the ratio of available faculty members to enrolled students. The higher this ratio, the more time will be available for faculty members to engage in research activities besides their teaching responsibilities. Research projects are represented as an aging-chain structure that mimics its lifecycle: proposals are submitted, some are granted, and the projects are executed and finalized. According to the faculty members' population profile, they can be related to sustainability or not.

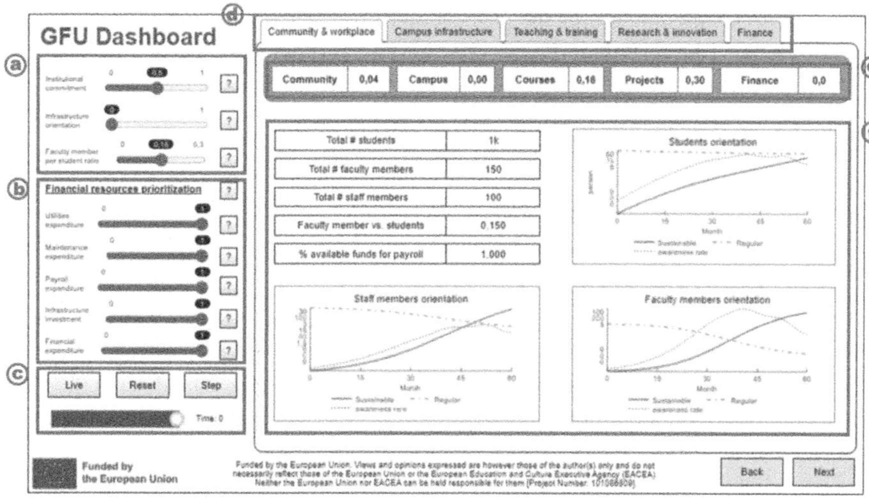

Fig. 4. Dashboard presenting KPIs, some of the simulation controls, and outputs.

Figure 4 identifies six sections of the simulation model's dashboard, which are:

a) *HEI's general setting*: Includes three sliders that control the level of commitment of HEI's leaders towards sustainability practices, and it influences the sustainable development adoption rate of the overall university populations, the infrastructure orientation indicating the proportion of a university's infrastructure investments allocated to sustainable (green) projects, and the faculty member per student ratio that reflects the level of personalized instruction and academic support available to students.
b) *Financial resources prioritization*: The users decide how they would like to allocate and prioritize fractions of the necessary operational expenditures and investments based on resource availability. A value between 0 and 1 imposes restrictions on each necessary expenditure category or investment, and restricting one category would increase the availability of resources for other necessities.
c) *Runtime controls*: Three buttons that make the model go live, where each change to any parameter immediately reflects in the simulation output from time zero to the end of the simulation time horizon, to run step by step, and the user can change the value of the parameter in each step, and the reset that clen up the run output and set the parameters' value to the initial condition.
d) *Subsystems navigation menu*: A tabbed upper menu allows the user to navigate through information on the four subsystems (Community & workplace, Campus infrastructure, Teaching & training, and Research & innovation) and financial information depicting the available financial resources that impose constraints to daily operations and necessary investments.
e) *Main KPIs*: There is a KPI for each of the four subsystems and one for the HEI's overall financial health. Those KPIs include the *Sustainability Awareness Ratio (SAR)*, which measures the proportion of the campus population who are aware of sustainability; the *Sustainable Infrastructure Ratio (SIR)*, which measures the proportion of green-oriented campus infrastructure and facilities; the *Sustainability Education*

Ratio (SER), which measures the proportion of regular courses that include sustainability topics; the *Sustainability Research Ratio (SRR)* measures the proportion of ongoing research projects related to sustainability topics; and the *Debt to Cash-Flow Ratio (DCFR)* indicates how many years it would take for an institution to repay its total debt using its current operating cash flow

f) *Additional indicators*: After selecting the desired subsystem through the menu identified as section "d," the user is presented with additional information about the selected subsystem. This information is presented as single values, plots with time series data, and other types of graphical representation.

The developed simulation model was used as a support tool for the project's participating HEIs to build their whole institution sustainability plans by revising the mental models of the participating team and stakeholders involved to understand better the intertwined relations of the multifaceted and multidimensional complex challenges involved in the green transition.

5 Conclusion and Final Remarks

Sustainability is a complex system transition that requires a holistic approach. Higher education institutions (HEIs) are complex systems with many interconnected parts, and changing one part can have ripple effects throughout the system. This makes it challenging to implement sustainability initiatives without considering the broader system.

The transition to sustainability in HEIs is a complex and multifaceted challenge that requires systemic approaches. This study presents an interactive system dynamics model to support HEIs in promoting awareness, strategic planning, and decision-making during the green transition. By integrating system dynamics with a whole-institution approach, the model provides a structured means to explore the relationships between policy decisions, institutional structures, and sustainability outcomes.

The developed simulation model focuses on key subsystems—community and workplace, campus infrastructure, teaching and training, and research and innovation—offering a dynamic and systemic representation of sustainable development processes within HEIs.

One of the challenges in transitioning to sustainability in higher education is the lack of alignment between the different parts of the system. For example, the curriculum may teach sustainability principles, but the institution's operational practices may not align with those principles, creating confusion and conflict for students, staff, and faculty.

Another challenge is the lack of resources. Sustainability initiatives can be expensive, and many higher education institutions face budget constraints. This can make it challenging to implement sustainability initiatives without sacrificing other priorities. At the same time, compliance with regulatory frameworks can create barriers that constrain the options available to institutions. Such frameworks can vary on a national basis, making systematic comparisons across national borders difficult and, thus, complicating the adoption of existing good practices.

Future research should focus on empirical testing by comparing model outcomes with actual institutional data and refining its assumptions. Furthermore, the model's scope should include broader policy contexts, such as national and international sustainability regulations, funding mechanisms, and benchmarking frameworks.

In conclusion, this research contributes to decision support in sustainability transitions by offering an interactive tool that bridges the gap between conceptual sustainability strategies and their practical implementation in HEIs. As sustainability challenges evolve, leveraging dynamic simulation approaches can give decision-makers the insights to drive meaningful, long-term transformation.

Acknowledgments. This study was funded by the European Union (project number 101086809). Views and opinions expressed are however those of the author(s) only and do not necessarily reflect those of the European Union or the European Education and Culture Executive Agency (EACEA). Neither the European Union nor EACEA can be held responsible for them.

Disclosure of Interests. The authors have no competing interests to declare that are relevant to the content of this article.

References

1. Brundtland, G.H.: Our Common Future: Report of the World Commission on Environment and Development. Geneva, UN-Dokument A/42/427 (1987). http://www.un-documents.net/ocf-ov.htm
2. Christou, O., Manou, D.B., Armenia, S., Franco, E., Blouchoutzi, A., Papathanasiou, J.: Fostering a whole-institution approach to sustainability through systems thinking: an analysis of the state-of-the-art in sustainability integration in higher education institutions. Sustainability **16**(6), 2508 (2024). https://doi.org/10.3390/su16062508
3. UNESCO: Issues and trends in education for sustainable development. In: Leicht, A., Heiss, J., Byun W.J. (eds.) UNESCO (2018). https://doi.org/10.54675/yelo2332
4. UNESCO: UNESCO roadmap for implementing the Global Action Programme on Education for Sustainable Development. UNESCO (2014)
5. Stöber, H., Gaebel, M., Morrisroe, A.: Greening in European higher education institutions. EUA Survey Data. European University Association (EUA) (2021)
6. UNESCO: Global citizenship education: preparing learners for the challenges of the 21st century. UNESCO (2014)
7. Armenia, S., Ferreira Franco, E., Mecella, M., Onori, R.: Smart model-based governance: from big-data to future policy making. In: Model-based Governance for Smart Organizational Future: BSLab-SYDIC International Workshop, Rome, January, pp. 23–24 (2017)
8. European Commission: The European Green Deal. COM(2019) 640 (2019)
9. European Commission: Directorate-General for Communication, European green deal – Delivering on our targets, Publications Office of the European Union (2021)
10. Bertalanffy, L.V.: General system theory: Foundations, development, applications. G. Braziller, New York (1968)
11. Schwaninger, M.: Intelligent Organizations - Powerful Models for Systemic Management. Springer, Heidelberg (2009). https://doi.org/10.1007/978-3-540-85162-2
12. Senge, P.M.: The Fifth Discipline: The Art & Practice of the Learning Organization. Image Books, New York (2006)
13. Meadows, D.H.: Thinking in Systems: A Primer. Chelsea Green Publishing, White River Junction (2008)

14. Forrester, J.: Industrial Dynamics, 1st edn. The MIT Press, Cambridge (1961)
15. Forrester, J.: Urban Dynamics, 1st edn. The MIT Press, Cambridge (1962)
16. Forrester, J.: World Dynamics, 1st edn. The MIT Press, Cambridge (1963)
17. Sterman, J.: Business Dynamics: Systems Thinking and Modeling for a Complex World. McGraw-Hill/Irwin, Boston (2000)
18. Blizzard, J.L., Klotz, L.E.: A framework for sustainable whole systems design. Des. Stud. **33**(5), 456–479 (2012). https://doi.org/10.1016/j.destud.2012.03.001
19. Menon, S., Suresh, M.: Synergizing education, research, campus operations, and community engagements towards sustainability in higher education: a literature review. Int. J. Sustain. High. Educ. **21**(5), 1015–1051 (2020). https://doi.org/10.1108/ijshe-03-2020-0089
20. Weiss, M., Barth, M.: Global research landscape of sustainability curricula implementation in higher education. Int. J. Sustain. High. Educ. **20**(4), 570–589 (2019). https://doi.org/10.1108/ijshe-10-2018-0190
21. Hargis, K., McKenzie, M., LeVert-Chiasson, I.: A whole institution approach to climate change education: preparing school systems to be climate proactive. Curriculum and learning for climate action. Brill **2021**, 43–66. (2021). https://doi.org/10.1163/9789004471818_004
22. Kohl, K., et al.: A whole-institution approach towards sustainability: a crucial aspect of higher education's individual and collective engagement with the SDGs and beyond (2021). https://doi.org/10.1108/ijshe-10-2020-0398
23. Chow, A.: One educational technology colleague's journey from dotcom leadership to university e-learning systems leadership: merging design principles, systemic change and leadership thinking. TechTrends **57**(5), 64–73 (2013). https://doi.org/10.1007/s11528-013-0693-6
24. Leal Filho, W., et al.: Sustainable Development Goals and sustainability teaching at universities: falling behind or getting ahead of the pack? J. Clean. Prod. **232**, 285–294 (2019). https://doi.org/10.1016/j.jclepro.2019.05.309
25. Lozano, R., et al.: A review of commitment and implementation of sustainable development in higher education: results from a worldwide survey. J. Clean. Prod. **108**, 1–18 (2015). https://doi.org/10.1016/j.jclepro.2014.09.048
26. Vennix, J.A.M.: Group Model Building: Facilitating Team Learning Using System Dynamics. Wiley, Chichester (1996)
27. Vennix, J.A., Akkermans, H.A., Rouwette, E.A.: Group model-building to facilitate organizational change: an exploratory study. Syst. Dyn. Rev. **12**(1), 39–58 (1996). 10.1002/(SICI)1099-1727(199621)12:1%3C39::AID-SDR94%3E3.0.CO;2-K
28. Hovmand, P.S.: Community Based System Dynamics. Springer, New York (2014)
29. iSee Systems. Stella Architect. https://www.iseesystems.com/store/products/stella-architect.asp. Accessed 20 Feb 2025

From Technical Debt to Business Process Debt: A Framework for Proactive Debt Management in BPM

Nikolaos Nousias, George Tsakalidis, George Nedos, and Kostas Vergidis(✉)

Department of Applied Informatics, University of Macedonia, Thessaloniki, Greece
{nnousiasuom,gtsakalidisuom,mai22049uom,kvergidisuom}@edu.gr

Abstract. Business Process Management (BPM) is a critical methodology for organizations aiming to enhance operational efficiency and achieve strategic objectives. While BPM's iterative lifecycle approach ensures continuous improvement, it is prone to shortcuts and compromises that prioritize short-term benefits over long-term sustainability. This tradeoff has been previously examined by the authors and is referred to as Business Process Debt (BPD), akin to Technical Debt (TD) in software engineering. However, unlike TD, which benefits from a well-defined set of indicators to identify its presence, the concept of BPD lacks a corresponding framework or established indicators. This paper addresses this gap by proposing a framework to identify potential BPD specifically during the business process modeling phase. Drawing parallels from software metrics used to define TD indicators, the framework leverages validated business process model metrics correlated with the External Quality Characteristic (EQC) of maintainability. The framework introduces the Total Maintainability Score (TMS), which aggregates these metrics to evaluate the overall maintainability of business process models. By classifying models into low, moderate, or high maintainability levels, and recognizing the inverse relationship between maintainability and debt, the TMS provides a practical indicator of potential BPD. The framework's applicability is demonstrated through a case study of the European Commission's New Computerized Transit System (NCTS). The findings highlight its potential to facilitate informed decision-making and proactive BPD management, thereby supporting long-term sustainable BPM initiatives.

Keywords: Business Process Debt · BPM · BPMN · Process Complexity · External Quality Characteristics · Decision-making

1 Introduction

Achieving operational efficiency and strategic objectives in today's dynamic and competitive corporate environment relies heavily on effective Business Process Management (BPM) [1]. BPM encompasses methods and tools designed to streamline processes, reduce costs, and improve performance, serving as a critical foundation for organizations aiming to optimize their operations. It is typically structured as a lifecycle methodology, integrating process improvement into an iterative feedback cycle [2]. This approach

B. Delibašić et al. (Eds.): ICDSST 2025, LNBIP 546, pp. 139–154, 2025.
https://doi.org/10.1007/978-3-031-90863-7_10

underscores the fundamental BPM principle of continuity [3], emphasizing that BPM should be considered a continuous practice rather than a one-time project [4].

While BPM offers significant benefits, its success is often subject to various pitfalls [5]. A common challenge arises when organizations take shortcuts or make compromises during the BPM lifecycle, prioritizing short-term gains over long-term sustainability. This can lead to what is known as Business Process Debt (BPD) [6]. Similar to Technical Debt (TD) in software engineering, BPD refers to shortcuts, compromises, or inefficiencies in business process management undertaken to achieve short-term benefits, which ultimately require additional effort to resolve in the future. Like TD, which can be incurred throughout the entire Software Development Lifecycle (SDLC), BPD can emerge at any phase of the BPM lifecycle, highlighting the importance of proactively identifying potential instances of BPD.

However, unlike TD - which is supported by a well-established set of indicators for identifying its presence - the literature currently lacks a framework or set of indicators for BPD. To address this gap, this paper proposes a framework to identify potential BPD specifically during the modeling phase of the BPM lifecycle. This focus is justified for the following reasons: (i) Business process modeling represents the most critical phase in the BPM lifecycle [7]. (ii) Any shortcuts or inefficiencies introduced during the modeling phase are likely to cascade into subsequent phases, leading to costly redesign initiatives later. (iii) The introduction of BPD indicators during the modeling phase can assist in identifying and mitigating potential inefficiencies early in the process.

In line with this direction and drawing parallels from the use of software metrics to define TD indicators, the authors utilize previously validated business process model metrics to inform the identification of BPD. These metrics have been correlated in prior research [8] with the External Quality Characteristic (EQC) of Maintainability, which is prioritized in this study due to its strong relationship with BPD. Specifically, a business process model with low maintainability is likely to require significant additional effort in the future to modify and preserve its quality, signaling a high potential for BPD. In contrast, a model with high maintainability suggests low potential BPD, as it allows process modelers to more easily adapt and maintain the model as the process evolves. However, since no single metric can fully evaluate a quality characteristic [8], such as maintainability, the authors introduce the Total Maintainability Score (TMS), which aggregates the selected metrics to provide a comprehensive maintainability assessment. Based on predefined TMS ranges, potential BPD can be identified.

Overall, the main contribution of this paper is the proposition of a framework to identify potential BPD during the business process modeling phase of the BPM lifecycle. By introducing a BPD indicator in modeling initiatives, practitioners are expected to identify and mitigate potential inefficiencies from their outset, thereby facilitating informed decision-making and overall BPD management.

The remainder of this paper is organized as follows: Sect. 2 establishes the background for this research, focusing on the key concepts of Technical Debt (TD), Business Process Debt (BPD), Business Process Model Metrics, and decision-making in debt management. Section 3 outlines the proposed framework and introduces the Total Maintainability Score (TMS). To demonstrate the framework's applicability, Sect. 4 presents a case study involving the European Commission's New Computerized Transit System

(NCTS). Finally, Sect. 5 concludes the paper by discussing the key findings, practical implications, limitations, and directions for future work.

2 Background

This section introduces the key concepts that form the foundation of the paper. First, the TD metaphor is explored, including its occurrence across various phases of the SDLC and its associated indicators. Next, the concept of BPD is presented, followed by a discussion on the rationale for leveraging process model metrics related to the maintainability EQC as the basis for the proposed framework. Finally, the section concludes with the decision-making process for debt management and the role of debt indicators within it.

2.1 Technical Debt (TD)

In science and various disciplines, metaphors play a crucial role in explaining abstract phenomena, making reasoning and communication more intuitive [9]. One such metaphor, widely used in software engineering, is the concept of TD. Introduced by Ward Cunningham in 1992 [10], the TD metaphor helps non-technical stakeholders understand the long-term costs associated with taking shortcuts or making compromises during software development [11]. It illustrates how such decisions, while beneficial in the short term, can potentially lead to increased complexity, effort, and costs over time if not addressed promptly [12]. In this way, TD serves both as a strategic tool for seizing immediate opportunities (e.g., entering the market early, gathering early feedback, etc.) and as a potential liability that accrues "interest" over time [13].

Although TD has traditionally been associated with code-related artifacts [14, 15], research has increasingly recognized that debt can arise throughout the entire SDLC [12, 14–18]. Alves et al. [18] proposed an ontology to systematically classify different types of TD based on their inherent nature, specifically the phase of the SDLC in which the debt originated or with which it is associated. For instance, when a developer opts for a suboptimal design to address an immediate issue, it may expedite progress in the short term but result in costly rework later, leading to design debt. Similarly, insufficient or missing testing coverage can be classified as testing debt, reflecting deferred efforts to ensure software quality.

Given that there is no universal standard for what constitutes a "good," "mature", or "complete" artifact (e.g., in implementation, design, testing, etc.), TD must be evaluated within the context of the system's future evolution and maintenance [19, 20]. For example, if a system contains poor-quality code but does not require future development or maintenance, it would not constitute TD. Consequently, only artifacts that might hinder future maintenance should be viewed as sources of debt, emphasizing the significant relationship between TD and system maintainability. However, since future maintenance is inherently unpredictable, these artifacts represent potential TD in the present. They only become effective TD when additional maintenance effort (e.g., higher cost, time) is eventually required [19].

Towards this direction, and to bring visibility to potential TD across different stages of the SDLC, previous research has introduced TD indicators. Among these, the most extensively studied is the concept of code smells [15]. Code smells refer to specific characteristics in source code that suggest potential issues or deviations from established best practices, such as principles of inheritance and encapsulation in object-oriented programming. Their presence often indicates areas where quality or design improvements have been deferred, signaling instances of potential TD as additional effort during future maintenance might be required [21]. Additionally, other studies [22, 23] have explored the relationship between software metrics and TD, proposing coupling, cohesion, and code duplication as further TD indicators.

2.2 Business Process Debt (BPD)

While the concept of TD has been extensively studied within the software development community, its principles are also relevant to other disciplines [24]. In this context, the authors of this study have previously applied the concept of TD to the domain of BPM [6], highlighting two key parallels: (i) prioritization and trade-offs are inevitable in BPM, much like in software development, and (ii) BPM success is often compromised by shortcuts taken during the BPM lifecycle, favoring short-term gains at the expense of long-term sustainability - directly contradicting the fundamental BPM principle of continuity [3]. Based on these parallels, the authors introduced the concept of BPD [6]. Analogous to TD, BPD refers to shortcuts, compromises, or inefficiencies in the management of business processes to achieve short-term gains, which accumulate over time and ultimately require additional effort to resolve.

Similar to the multiple sources of TD across the SDLC, BPD can be incurred at any stage of the BPM lifecycle. The BPM lifecycle follows a structured and iterative approach to managing and continuously improving business processes, typically consisting of the following phases: Specification, Modeling, Analysis, Implementation, Execution, and Evaluation [7]. Each phase is susceptible to shortcuts or suboptimal practices that contribute to the accumulation of BPD. For example, a quickly designed business process model that fails to account for process exceptions may lead to inefficient process implementation. Over time, this inefficiency may necessitate costly redesign, which can be classified as business process modeling debt.

2.3 Business Process Modeling and Metrics

Business process modeling is widely utilized by organizations to enhance understanding and awareness of their operations [25] and plays a critical role as a phase in the BPM lifecycle [7]. This activity involves documenting how businesses execute their processes, typically through the creation of a conceptual model that illustrates activities, events, and control flow logic [26]. The most commonly adopted notation for this purpose is Business Process Model and Notation (BPMN) [27]. Introduced by the Object Management Group (OMG) in 2006, BPMN has become the de facto standard in both academia and industry due to its formal yet accessible notation for diverse business users [28].

As the primary purpose of modeling is to facilitate communication, prior research has introduced a range of metrics to evaluate the quality of process models [29–31]. These metrics are primarily focused on the structural complexity of the model, offering critical insights into its internal quality. However, assessing a model's quality should not be limited to its internal properties; it must also account for its cognitive complexity and impact on users (i.e., external quality) [8]. To incorporate this broader perspective, Sánchez-González et al. [8] have previously correlated process model metrics with External Quality Characteristics (EQCs) such as maintainability, usability, and correctness.

In this study, metrics associated with maintainability are prioritized due to the well-established relationship between debt and maintainability. According to [8], maintainability is defined as the degree to which a model can be effectively and efficiently modified by its intended maintainers without introducing defects or degrading its existing quality. A business process model with low maintainability can signal high potential BPD, as additional effort may be required to modify and maintain the model. Conversely, a model with high maintainability indicates low potential BPD, as process modelers are more likely to effectively adjust and maintain the model over time.

Therefore, similar to how software metrics serve as quantitative measures to provide insights into the quality of software artifacts and the presence of TD, this study employs process model metrics correlated with the maintainability EQC (Table 1) as the foundation for identifying potential instances of BPD.

2.4 Decision-Making and Debt Management

The primary objective of identifying potential debt (whether it is TD or BPD) is to enhance and support the decision-making process for effective debt management [32].

According to prior research [32, 33], effective debt management requires informed decision-making at two key stages. Initially, an explicit decision should be made on whether to incur debt. Second, once debt is incurred, a critical decision involves determining whether, when, and how to repay it. These decision-making processes are inherently complex because they involve balancing immediate benefits against the future costs associated with maintenance or refactoring. The underlying decision logic centers on optimizing the tradeoff between cost and value [14]. To support this decision-making, prior research has adopted methodologies from finance, such as cost-benefit analysis and portfolio management, alongside approaches from decision science literature, such as the Analytic Hierarchy Process (AHP) [32].

A fundamental prerequisite for both decision-making processes is having visibility into the level of debt. Without such visibility, decisions are often driven by project constraints, such as time pressures, rather than a thorough consideration of the long-term tradeoffs associated with maintenance [34]. This lack of transparency can lead to shortcuts and deferred actions, which accumulate debt over time and ultimately shift the focus toward reactive debt management.

To address this challenge, the next section introduces a proposed framework for identifying potential BPD in business process modeling initiatives, aiming to facilitate informed decision-making and support proactive BPD management.

3 Proposed Framework

This section introduces the proposed framework for identifying potential BPD in modeling initiatives. The framework phases are first described, followed by an explanation of the methodology for calculating and interpreting the Total Maintainability Score (TMS).

3.1 Overview of the Framework Phases

The proposed framework is illustrated in Fig. 1, highlighting its alignment with the existing literature, its structured approach to identifying BPD (i.e., its phases), and its implications in practice (refer to Discussion). The framework consists of four sequential phases which are consequently presented in detail.

Metric Calculation
The first phase involves calculating a predefined set of process model metrics (Table 1) that are correlated with the External Quality Characteristic (EQC) of maintainability. These metrics were derived from prior research [8] and their calculation is based upon inspecting the structural elements of the process model (e.g., nodes, gateways, flows, etc.), providing insight into its structural complexity (e.g., model's size, linearity, control-flow complexity, etc.).

Complexity Level Assessment
The second phase involves assessing the complexity level of each previously calculated process model metric. Using thresholds derived from prior research [31, 35], each metric is categorized into one of three levels: Low, Moderate, or High Complexity (Table 1). These thresholds are metric-specific, ensuring an objective and accurate classification of complexity.

Aggregation into Total Maintainability Score (TMS)
In the next phase, the complexity levels of all metrics are aggregated into a single measure, referred to as the Total Maintainability Score (TMS) (refer to Subsect. 3.2). Introduced by the authors, the TMS serves as a novel approach to synthesizing the complexity levels of multiple metrics into a unified evaluation of maintainability. This is because quality characteristics are inherently multifaceted and cannot be adequately assessed using only a single measure or a subset of metrics [8]. By incorporating all metrics, the framework accounts for a broader range of structural properties that influence maintainability, such as process flow complexity, model size, and interconnections.

TMS Interpretation and BPD Identification
In the final phase, the TMS is interpreted to assess whether the business process model exhibits indications of potential BPD. Based on predefined ranges for the TMS (Subsect. 3.3), the overall maintainability level is categorized, and corresponding potential BPD levels are identified.

Table 1. Business process model metrics related to maintainability ECQ and their thresholds to define complexity level

Metric	Full Metric Title	Description	Complexity Thresholds		
			Low	Moderate	High
AGD	Average Gateway Degree	Average number of incoming and outgoing flows for all gateways in a process model.	≤3.83	3.83 < AGD ≤ 4.06	>4.06
CFC	Control Flow Complexity	Complexity of the process's control flow based on the XOR, OR, AND split structures present in a process.	≤12.1	12.1 < CFC ≤ 26.9	>26.9
Depth	Depth	Longest sequence of steps from the start to the end event in a process model, providing a measure of the structural complexity	≤1.56	1.56 < Depth ≤ 3.26	>3.26
GH	Gateway Heterogeneity	Variety of different types of gateways (e.g., XOR, AND, OR) in a process model.	≤7.15	7.15 < GH ≤ 13.2	>13.2
GM	Gateway Mismatch	Alignment between gateway splits and joins in a process, measuring instances where the type of a splitting gateway (e.g., XOR, AND) does not match its corresponding joining gateway.	≤12	12 < GM ≤ 24	>24

(*continued*)

Table 1. (*continued*)

Metric	Full Metric Title	Description	Complexity Thresholds		
			Low	Moderate	High
MGD	Maximum Gateway Degree	Highest number of incoming and outgoing flows associated with a single gateway in the model, indicating the most complex decision point in the process.	≤5	5 < MGD ≤ 7	>7
NSFG	Number of Sequence Flows from Gateways	Total number of sequence flows from all gateways, indicating the model's linearity or potential complexity.	≤9.26	9.26 < NSFG ≤ 18.4	>18.4
# of Nodes	Number of Nodes	Total number of nodes (activities, events, gateways, etc.) in a process model, providing a basic measure of the model's size and complexity.	≤29.9	29.9 < #Node ≤ 58.1	>58.1
Separability	Separability	The degree (%) to which a process can be divided into independent subprocesses.	≥0.68	0.68 < Sep. ≤ 0.42	<0.42
TNG	Total Number of Gateways	Total number of gateways in a process model, reflecting the complexity of decision-making and routing within the process.	≤12	12 < TNG ≤ 16	>16

3.2 Calculation of the Total Maintainability Score (TMS)

To assess the total maintainability of a business process model, the authors introduce the Total Maintainability Score (TMS) (1):

$$TMS = \frac{\sum_{i=1}^{n} (Wi\ Ci)}{\sum_{i=1}^{n} (Wi\ Cmax)} \tag{1}$$

In this equation, n denotes the total number of evaluated metrics (i.e., here $n = 10$, refer to Table 1), Wi represents the weight assigned to the i-th metric, Ci is the observed complexity value for the i-th metric, and $Cmax$ signifies the maximum possible complexity value for any given metric.

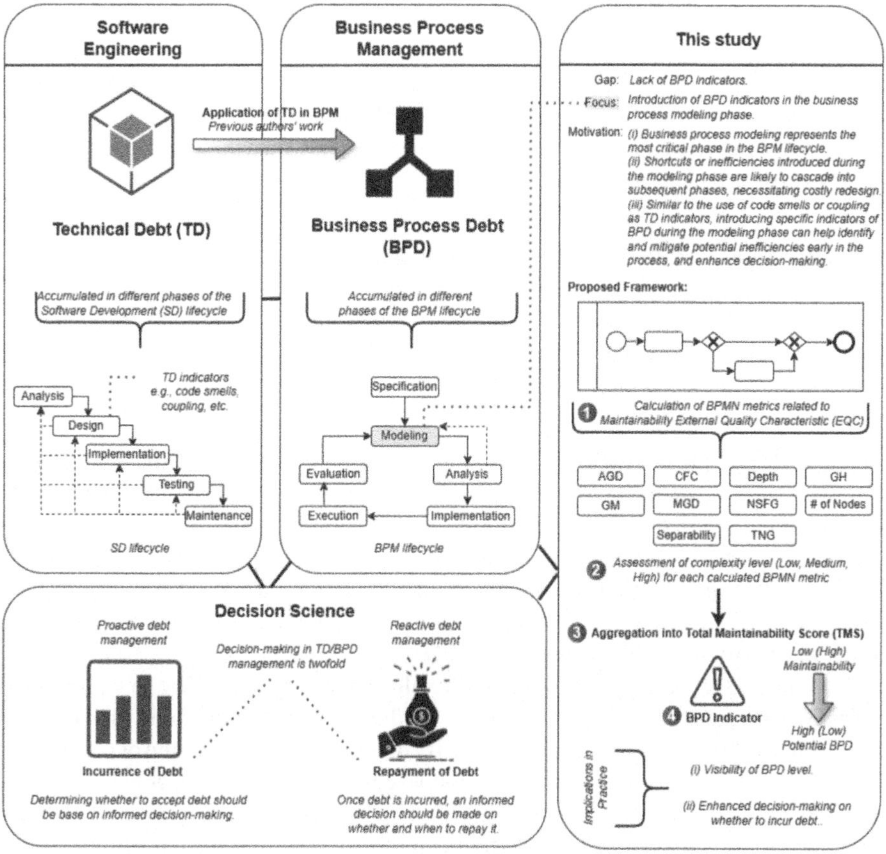

Fig. 1. Proposed framework for identifying potential BPD in business process modeling

The value of Ci is determined based on the complexity level of the evaluated metric: If a metric demonstrates high complexity, its Ci value is set to 3, signifying a significant contribution to the overall complexity of the process model. If a metric exhibits moderate complexity, its Ci value is 2, reflecting a balanced level of complexity. For metrics with low complexity, Ci is assigned a value of 1, indicating minimal impact on the complexity of the model. This tiered assignment of Ci values ensures that metrics with higher levels of complexity have a proportionally greater influence on the TMS calculation.

The $Cmax$ value is consistently set to 3, representing the highest possible complexity level for any given metric. This standardization allows the formula to normalize the weighted sum of observed complexity values, ensuring that the TMS always lies within the range of 0 (indicating maximum maintainability - Subsect. 3.3) to 1 (indicating minimum maintainability - Subsect. 3.3). The rationale for setting $Cmax$ to 3 aligns with the predefined scale of complexity levels (low, moderate, and high), where high complexity is recognized as the upper bound of the proposed framework.

In the current approach, the authors have considered all metrics as equally significant, assigning a uniform weight of $Wi = 1$ for all metrics. This decision reflects the assumption that each metric contributes equally to the overall maintainability of the process model. This simplification is particularly useful when no empirical evidence is available to suggest that certain metrics are inherently more critical than others. By assigning equal weights, the model avoids introducing bias and provides a baseline framework that can be adapted in future studies if differentiated weights are justified through further research or domain-specific considerations.

3.3 TMS Interpretation and BPD Identification

To interpret the TMS effectively, the authors establish specific thresholds:

- High Maintainability: $0.0 \leq TMS \leq 0.3$
- Moderate Maintainability: $0.3 < TMS \leq 0.6$
- Low Maintainability: $0.6 < TMS \leq 1.0$

These thresholds categorize process models into three levels of maintainability: Low Maintainability accounts for 40% of the scale, while Moderate and High Maintainability each represent 30%. This distribution reflects the authors' intention to adopt a stricter evaluation criterion for identifying potential BPD. By extending the range for Low Maintainability, the framework ensures that borderline cases - those with TMS values slightly above 0.6 - are classified as low maintainability, thereby prompting closer scrutiny and earlier intervention to mitigate potential BPD.

The inverse relationship between TMS values and maintainability (e.g., TMS 0.2 indicates high maintainability, while TMS 0.8 indicates low maintainability) is a result of the formula's (1) normalization mechanism. By dividing the weighted sum of actual complexities $\sum_{i=1}^{n}(WiCi)$ by the maximum possible complexity $\sum_{i=1}^{n}(WiCmax)$, , the score becomes a proportional indicator. Higher TMS values indicate that a process exhibits a significant share of its potential maximum complexity (i.e., low maintainability), while lower values reflect a smaller share, correlating directly with better maintainability.

Building on this observation, the TMS serves as an indicator of potential BPD. Specifically, low maintainability levels ($TMS > 0.6$) points to a process model with significant structural complexity and maintenance challenges, indicating high potential BPD. Such models are difficult to understand and adapt and are likely to eventually require costly redesign efforts to address their inefficiencies. Conversely, high maintainability levels

($TMS \leq 0.3$) reflects a well-structured and easily maintainable process model, suggesting low potential BPD. These models are aligned with best practices and modeling guidelines, minimizing inefficiencies, and reducing the likelihood of requiring significant future redesign efforts. Finally, moderate maintainability levels ($0.3 < TMS \leq 0.6$) represent a balanced state, where the risk of BPD is neither negligible nor critical. While models in this category are generally manageable, they require ongoing monitoring to ensure that their maintainability does not degrade over time, potentially increasing the risk of BPD.

To support these interpretations, a graphical representation of TMS values and their relationship with potential BPD is provided in Fig. 2. The figure illustrates an example where a TMS value of 0.8 signifies low maintainability of the business process model, thereby indicating high potential BPD.

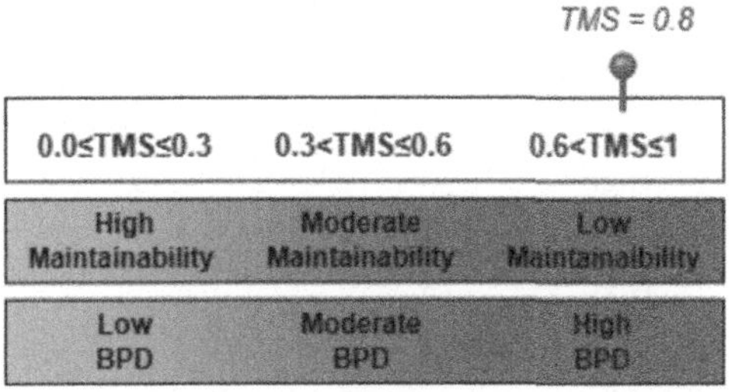

Fig. 2. Interpretation of TMS and identification of potential BPD

4 Case Study: European Commission's New Computerized Transit System (NCTS)

To validate the proposed framework, this study applies its methodology to the European Commission's New Computerized Transit System (NCTS), Phase 5 [36]. The NCTS is a key initiative by the European Commission designed to modernize and streamline customs transit procedures across the European Union. The system aims to enhance transparency, efficiency, and compliance with EU and international standards by integrating various stakeholders (i.e., including customs authorities, businesses, and logistics providers) operating under diverse regulatory frameworks.

The case study involves the analysis of 96 BPMN models, the calculation of metrics, and the identification of the overall potential level of BPD across all models collectively. These models represent intricate workflows of the system, encompassing operational interdependencies, control flows, and decision-making processes.

4.1 Calculation of Metrics and TMS

To determine the overall potential level of BPD, the proposed framework was applied systematically. First, the predefined set of process model metrics (see Table 1) was calculated for each of the 96 BPMN models. Since the objective was to evaluate the overall potential level of BPD, the average value of each metric across all models was calculated (refer to the Average column in Table 2). In the next phase of the framework, the complexity level for each metric was determined based on the predefined thresholds (see Table 1), with the corresponding values recorded in the Complexity column of Table 2. These complexity levels were then aggregated into the TMS in the subsequent step. Specifically, the observed complexity *Ci* for each metric was calculated, and the weighted sum of these values was determined, resulting in a total score of 23. Given that the maximum complexity level for any metric is 3 and that 10 metrics were assessed, the maximum possible score was 30. By dividing the weighted score by the maximum possible score, the TMS value was obtained.

Table 2. Calculation of Metrics and TMS

Metric	Average	Complexity	Ci	Weighted Score	Maximum Score	TMS	Potential BPD
AGD	4.75	High	3	23	30	0.77	High
CFC	21.5	Moderate	2				
Depth	5.26	High	3				
GH	2.78	Low	1				
GM	2.34	Low	1				
MGD	7.26	High	3				
NSFG	21.56	High	3				
# of Nodes	38.05	Moderate	2				
Separability	0.29	High	3				
TNG	13.01	Moderate	2				

4.2 Results

The calculated TMS of 0.77 indicates a high overall complexity in the analyzed BPMN models, suggesting significant challenges in process maintainability. This high complexity corresponds to a high potential for BPD, which could manifest as effective BPD if modifications or further evolution of these models are required in the future.

This complexity is primarily driven by the excessive use of gateways, as reflected in the high complexity levels of AGD (4.75), MGD (7.26), and NSFG (21.56). These metrics demonstrate that the models incorporate many decision points and multiple outgoing flows per gateway, increasing structural intricacy. Furthermore, the Depth metric (5.26) also falls within the high complexity range, indicating that the longest execution path in the models is relatively deep. This suggests that the processes involve multiple sequential steps before reaching completion, potentially reducing their flexibility and maintainability. Additionally, Separability (0.29) is categorized as high complexity, reinforcing the notion that the processes are not easily decomposable into independent subprocesses, further impacting modularity and maintainability. Conversely, some metrics exhibit lower complexity levels, providing a degree of balance. GH (2.78) and GM (2.34) are classified as low complexity, indicating that the models predominantly utilize a limited variety of gateway types and maintain consistent split-join relationships. Finally, CFC (21.5), Number of Nodes (38.05), and TNG (13.01) fall within the moderate complexity range, suggesting that while these factors contribute to overall maintainability concerns, they do not reach excessive levels of complexity.

Overall, the results indicate that gateway overuse and a low degree of process decomposition are the primary contributors to the high complexity observed in the models, directly impacting maintainability. This suggests that optimizing gateway usage and enhancing process modularity could improve maintainability. These findings align with existing literature, which emphasizes the importance of reducing unnecessary decision points [37] and enforcing modularity [38] in process models to enhance their long-term maintainability.

5 Discussion and Conclusions

This paper proposes a structured framework for identifying BPD during the business process modeling phase of the BPM lifecycle, a topic that has been unexplored in existing literature. The study leverages validated process model metrics related to the EQC of maintainability and introduces the TMS metric to assess the overall maintainability of business process models. This approach offers a comprehensive methodology for identifying potential inefficiencies in process models, thereby providing a timely contribution towards the indication of potential BPD. The key findings and implications of this study are outlined below:

(i) BPD operates similarly to other forms of debt: Shortcuts and compromises made during the modeling phase may deliver short-term benefits at the expense of accumulating inefficiencies and increased future costs, such as higher maintenance efforts and costly redesign. This highlights the critical need for balancing speed with long-term sustainability in business process design. For practitioners, this finding emphasizes the importance of adhering to best practices and employing modeling strategies that prevent debt accumulation, ensuring efficient and maintainable processes over time.

(ii) Maintainability bears an inverse relationship with BPD: A business process model with low maintainability indicates a higher likelihood of BPD, as more effort is required to sustain and adapt the process over time. Conversely, process models with high maintainability exhibit reduced potential for BPD. Such models are typically aligned with existing best practices and modeling guidelines. For practitioners, this finding reinforces the necessity of mitigating structural inefficiencies in business process models to minimize the accumulation of BPD.

(iii) Visibility of BPD enables proactive debt management: Identifying potential BPD in business process modeling allows organizations to detect inefficiencies early in the BPM lifecycle. Early detection creates opportunities for timely interventions, preventing these issues from escalating in later phases and avoiding costly redesigns. This proactive approach shifts organizations away from reactive debt management, allowing them to address inefficiencies before they accumulate. For practitioners, BPD visibility is particularly valuable, as it enables informed decision-making, ensuring that any BPD incurred is the result of strategic choices rather than unintended actions or project constraints.

While this study represents an initial step towards the effective management of BPD, several limitations should be acknowledged. First, the applicability and validation of the proposed framework rely on preliminary results on a case study of the European Commission's NCTS. Although this case provides a practical context, its unique characteristics may limit the generalizability of the findings to other organizational contexts or industries. Second, the framework assumes equal weights for all metrics when calculating the TMS. While this simplifies the analysis, it may fail to account for scenarios where specific metrics have a disproportionate impact on maintainability and, consequently, on BPD. Finally, although the thresholds applied to the metrics are derived from established literature, further empirical validation across diverse datasets is necessary to refine these thresholds and improve their applicability within the proposed framework.

As future work, several areas for further research are proposed. As the calculation of metrics constitutes a manual process that requires time, one promising direction involves the development of a tool capable of automatically calculating the TMS score and signaling potential BPD. Such a tool would assess the structural elements of a process model in real-time as the process is being designed, calculate the corresponding metrics, and present the TMS score alongside the identified level of potential BPD. Additionally, future research could focus on validating the framework across a broader range of domains and refining metric thresholds to improve their applicability in varied organizational settings. Expanding the scope of the framework to include adaptive thresholds and dynamic weighting mechanisms could further enhance its precision and relevance. Finally, future research could apply concepts from decision science to developing tools and strategies that support stakeholders in decision-making of balancing short-term gains with long-term sustainability in process design.

References

1. Tsakalidis, G., Vergidis, K.: A roadmap to critical redesign choices that increase the robustness of business process redesign initiatives. J. Open Innov. Technol. Market Complex. **7**, 178 (2021). https://doi.org/10.3390/joitmc7030178
2. Segatto, M., de Pádua, S.I.D., Martinelli, D.P.: Business process management: a systemic approach? Bus. Process. Manag. J. **19**, 698–714 (2013). https://doi.org/10.1108/BPMJ-Jun-2012-0064
3. Castro, B.K. do A., Dresch, A., Veit, D.R.: Key critical success factors of BPM implementation: a theoretical and practical view. Bus. Process Manage. J. **26**, 239–256 (2019). https://doi.org/10.1108/BPMJ-09-2018-0272
4. vom Brocke, J., Schmiedel, T., Recker, J., Trkman, P., Mertens, W., Viaene, S.: Ten principles of good business process management. Bus. Process. Manag. J. **20**, 530–548 (2014). https://doi.org/10.1108/BPMJ-06-2013-0074
5. Malinova, M., Mendling, J.: Identifying do's and don'ts using the integrated business process management framework. Bus. Process. Manag. J. **24**, 882–899 (2018). https://doi.org/10.1108/BPMJ-10-2016-0214

6. Nousias, N., Nedos, G., Tsakalidis, G., Vergidis, K.: Mitigating business process debt with digital process twins. In: Decision Support System in an Uncertain World: The Contribution of Digital Twins (2023)
7. Nousias, N., Tsakalidis, G., Vergidis, K.: Not yet another BPM lifecycle: a synthesis of existing approaches using BPMN. Inf. Softw. Technol. **171**, 107471 (2024). https://doi.org/10.1016/j.infsof.2024.107471
8. Sánchez-González, L., García, F., Ruiz, F., Piattini, M.: Toward a quality framework for business process models. Int. J. Coop. Info. Syst. **22**, 1350003 (2013). https://doi.org/10.1142/S0218843013500032
9. Glucksberg, S., Keysar, B.: How metaphors work. In: Metaphor and Thought, 2nd edn., pp. 401–424. Cambridge University Press, New York (1993). https://doi.org/10.1017/CBO9781139173865.020
10. Cunningham, W.: The WyCash portfolio management system. SIGPLAN OOPS Mess. **4**, 29–30 (1992). https://doi.org/10.1145/157710.157715
11. Spínola, R.O., Vetrò, A., Zazworka, N., Seaman, C., Shull, F.: Investigating technical debt folklore: Shedding some light on technical debt opinion. In: 2013 4th International Workshop on Managing Technical Debt (MTD), pp. 1–7 (2013). https://doi.org/10.1109/MTD.2013.6608671
12. Brown, N., et al.: Managing technical debt in software-reliant systems. In: Proceedings of the FSE/SDP Workshop on Future of Software Engineering Research, Santa Fe New Mexico USA. ACM (2010). https://doi.org/10.1145/1882362.1882373
13. Ampatzoglou, A., Ampatzoglou, A., Chatzigeorgiou, A., Avgeriou, P.: The financial aspect of managing technical debt: a systematic literature review. Inf. Softw. Technol. **64**, 52–73 (2015). https://doi.org/10.1016/j.infsof.2015.04.001
14. Li, Z., Avgeriou, P., Liang, P.: A systematic mapping study on technical debt and its management. J. Syst. Softw. **101**, 193–220 (2015). https://doi.org/10.1016/j.jss.2014.12.027
15. Alves, N.S.R., Mendes, T.S., de Mendonça, M.G., Spínola, R.O., Shull, F., Seaman, C.: Identification and management of technical debt: a systematic mapping study. Inf. Softw. Technol. **70**, 100–121 (2016). https://doi.org/10.1016/j.infsof.2015.10.008
16. Guo, Y., et al.: Tracking technical debt — an exploratory case study. In: 2011 27th IEEE International Conference on Software Maintenance (ICSM), pp. 528–531 (2011). https://doi.org/10.1109/ICSM.2011.6080824
17. Seaman, C., Guo, Y.: Chapter 2 - measuring and monitoring technical debt. In: Zelkowitz, M.V. (ed.) Advances in Computers, pp. 25–46. Elsevier (2011). https://doi.org/10.1016/B978-0-12-385512-1.00002-5
18. Alves, N.S.R., Ribeiro, L.F., Caires, V., Mendes, T.S., Spínola, R.O.: Towards an ontology of terms on technical debt. In: 2014 Sixth International Workshop on Managing Technical Debt, pp. 1–7 (2014). https://doi.org/10.1109/MTD.2014.9
19. Schmid, K.: On the limits of the technical debt metaphor some guidance on going beyond. In: 2013 4th International Workshop on Managing Technical Debt (MTD), pp. 63–66 (2013). https://doi.org/10.1109/MTD.2013.6608681
20. Kruchten, P., Nord, R.L., Ozkaya, I.: Technical debt: from metaphor to theory and practice. IEEE Softw. **29**, 18–21 (2012). https://doi.org/10.1109/MS.2012.167
21. Izurieta, C., Vetrò, A., Zazworka, N., Cai, Y., Seaman, C., Shull, F.: Organizing the technical debt landscape. In: 2012 Third International Workshop on Managing Technical Debt (MTD), pp. 23–26 (2012). https://doi.org/10.1109/MTD.2012.6225995
22. MacCormack, A., Sturtevant, D.J.: Technical debt and system architecture: the impact of coupling on defect-related activity. J. Syst. Softw. **120**, 170–182 (2016). https://doi.org/10.1016/j.jss.2016.06.007
23. Marinescu, R.: Assessing technical debt by identifying design flaws in software systems. IBM J. Res. Dev. **56**, 9:1–9:13 (2012). https://doi.org/10.1147/JRD.2012.2204512

24. Allman, E.: Managing Technical Debt: shortcuts that save money and time today can cost you down the road. Queue **10**, 10–17 (2012). https://doi.org/10.1145/2168796.2168798
25. Indulska, M., Green, P., Recker, J., Rosemann, M.: Business process modeling: perceived benefits. In: Laender, A.H.F., Castano, S., Dayal, U., Casati, F., de Oliveira, J.P.M. (eds.) Conceptual Modeling - ER 2009, pp. 458–471. Springer, Heidelberg (2009). https://doi.org/10.1007/978-3-642-04840-1_34
26. Recker, J., Rosemann, M., Indulska, M., Green, P.: Business process modeling - a comparative analysis. J. Assoc. Inf. Syst. **10** (2009). https://doi.org/10.17705/1jais.00193
27. Object Management Group: Business Process Model and Notation. https://www.omg.org/spec/BPMN. Accessed 15 Jan 2025
28. Chinosi, M., Trombetta, A.: BPMN: an introduction to the standard. Comput. Stand. Interf. **34**, 124–134 (2012). https://doi.org/10.1016/j.csi.2011.06.002
29. Cardoso, J.: Process control-flow complexity metric: an empirical validation. In: 2006 IEEE International Conference on Services Computing (SCC 2006), pp. 167–173 (2006). https://doi.org/10.1109/SCC.2006.82
30. Gruhn, V., Laue, R.: Complexity metrics for business process models. Presented at the Business Information Systems – 9th International Conference on Business Information Systems (BIS 2006) (2006)
31. Mendling, J.: Metrics for Process Models: Empirical Foundations of Verification, Error Prediction, and Guidelines for Correctness. Springer, Heidelberg (2008)
32. Seaman, C., et al.: Using technical debt data in decision making: potential decision approaches. In: 2012 Third International Workshop on Managing Technical Debt (MTD), pp. 45–48 (2012). https://doi.org/10.1109/MTD.2012.6225999
33. Buschmann, F.: To pay or not to pay technical debt. IEEE Softw. **28**, 29–31 (2011). https://doi.org/10.1109/MS.2011.150
34. Tom, E., Aurum, A., Vidgen, R.: A consolidated understanding of technical debt. Presented at the European Conference on Information Systems (2012)
35. Sánchez-González, L., García, F., Ruiz, F.: Business Process Model Improvement based on Measurement Activities. https://alarcos.esi.uclm.es/bpmima/indicators.htm. Accessed 12 Jan 2025
36. Directorate-General for Taxation and Customs Union: New Computerised Transit System (NCTS): The deployment of the phase 5 is on track - European Commission. https://taxation-customs.ec.europa.eu/news/new-computerised-transit-system-ncts-deployment-phase-5-track-2024-12-20_en. Accessed 12 Jan 2025
37. Hasić, F., De Smedt, J., Vanthienen, J.: Augmenting processes with decision intelligence: principles for integrated modelling. Decis. Support Syst. **107**, 1–12 (2018). https://doi.org/10.1016/j.dss.2017.12.008
38. Reijers, H., Mendling, J.: Modularity in process models: review and effects. In: Dumas, M., Reichert, M., Shan, M.-C. (eds.) Business Process Management, pp. 20–35. Springer, Heidelberg (2008). https://doi.org/10.1007/978-3-540-85758-7_5

Author Index

B. Delibašić et al. (Eds.): ICDSST 2025, LNBIP 546, p. 155, 2025.
https://doi.org/10.1007/978-3-031-90863-7

The manufacturer's authorised representative in the EU is Springer Nature Customer Service Centre GmbH, Europaplatz 3, 69115 Heidelberg, Germany. If you have any concerns regarding our products, please contact ProductSafety@springernature.com

Printed and bound by CPI Group (UK) Ltd, Croydon, CR0 4YY
30/09/2026
02225576-0012